I0066905

A. T. Urquhart

Art. XIX.— On new species of Araneæ

A. T. Urquhart

Art. XIX.— On new species of Araneæ

ISBN/EAN: 9783741197963

Manufactured in Europe, USA, Canada, Australia, Japa

Cover: Foto ©berggeist007 / pixelio.de

Manufactured and distributed by brebook publishing software
(www.brebook.com)

A. T. Urquhart

Art. XIX.— On new species of Araneæ

ART. XIX.—*On New Species of Araneæ.*

By A. T. URQUHART, Corr. Mem. Royal Society of
Tasmania.

[*Read before the Auckland Institute, 3rd November, 1890.*]

Plate XXI.

Fam. DYSDIRIDÆ.

Gen. Ooxors, Templeton.

Oonops septem-cincta, sp. nov.   Plate XXI., fig. 1.

Femina.— Ceph.-th., long, 5; broad, 3. Abd., long, 5;
broad, 2.   Legs, 1, 2, 4, 3   13, 12·5, 11·5, 10·3 mm.

Cephalothorax bright-mahogany colour, fuscous clouding
about margin of cephalic region; hairs very short, sparse;
elongate-oval, prominently convex, roundly truncated; *clypeus* inclined forwards, depth equal to the greater diameter
and one-half of a centre eye; thoracic indentation slight,
normal grooves faint; profile-contour represents an even arch,
dips somewhat abruptly across ocular area.

Eyes of about equal size, oval, opalescent, posited on low
dark eminences, in three subcontiguous groups; centre pair
placed slightly in advance of hind-laterals, perceptibly more
distant from them than they are from anterior laterals, an interval rather surpassing their greater diameter.

Falces lake-black; few hairs; conical, gibbous at base in
front, outwardly inclined, rather longer than radial + digital
joints of palpus; breadth at base exceeds one-half length.

Maxillæ much dilated at insertion of palpi, spathulate,
more distinctly curved on superior side.

Labium suboval, sides apparently pressed inwards by
maxillæ; about half as long as the latter; organs red-mahogany colour, clouded with a deeper shade.

Sternum fulvous; elongate-oval, terminates above fourth
pair of coxal joints, prolonged beyond first pair for a distance
equal to three-fourths breadth of lip.

Legs yellowish-mahogany, moderately strong; hairs fine,
rather long; femora of first, second, and third pairs have 2
short black spines on fore-end, fourth 1 distal spine, 2 basal;
tibia of first leg, 2, 2, 2 beneath, 2 side spines; metatarsus,
double row, 4-6; tibial joint of second, 2, 2, 2; metatarsus,
2, 2, 2: tibia of third leg, 1, 1, 2; metatarsus, 2, 1, 3: fourth
leg, tibial joint, 1, 2; metatarsus, 2, 1, 4, or 3, 1, 4.   Superior
tarsal claws—First pair strong, well-curved, outer 11 comb-
teeth; inner claw, 12 long, 1 short basal tooth; inferior claw
short, sharply bent. 1 long curved tooth.

Palpi and legs concolorous; stout; pars humeralis, length about equal to cubital + radial joints together; cubital joint somewhat shorter than penultimate article; pars digitalis scarcely as long as two former joints together; hairs numerous; palpal claw short, stout, well curved, no teeth.

Abdomen elongate-oviform; hairs somewhat sparse, chiefly on base; pale-yellowish stone-colour, crossed by seven brownish-purple, recurved, arcuate bars, thickening somewhat in centre; basal bar narrowest and straightest, connected with second and third bars by a rather narrow median band, interval separating two latter bars less than one-half space between the second and first; posterior bars closer to one another, less pronounced; ventral region speckled, displays a wide median streak, dilated at spinners; broad band above vulva, shade lighter than dorsal marks. Spinners orange-colour. *Corpus vulvæ* orange-yellow; wide, convex, represents a segment of a circle, projects over the rima genitalis. Two pairs of spiracular openings, second not far from first pair, and not quite so distant from each other.

According to Thorell, Simon, and Cambridge, there appears to be some doubt as to whether the typical form *O. pulcher,* Templeton, possesses more than two stigmata; the Rev. O. P. Cambridge believed that he could discern four. In *O. septemcincta* the first pair of spiracular openings are easily discernible, the second not so, in fact scarcely visible, but when stretched the stigmata prove to be as large and well-developed as the first pair.

I am indebted to *Mr. T. Kirk, F.L.S.,* for this interesting example, which was contained in his collection from Wellington.

Fam. AGELENIDÆ.

Gen. TEGENARIA, Ltr.

Tegenaria arboricola, sp. nov.   Plate XXI., fig. 8.

Mas.—Ceph.-th., long, 4; wide, 3.   Abd., long, 5; wide, 2·4.   Legs, 1, 4, 2, 3 = 24, 21, 18·5, 16 mm.

Cephalothorax fulvous, dorsal band broad, lightly shaded with black, bifurcates forwards at fovea, leaving a narrow streak, tapers rapidly close to posterior median eyes; similar shading on lateral margins; border narrow, dark; glabrous; pars cephalica moderately convex, quadrate, facial index surpasses lateral by one-fourth; *clypeus* convex, depth equals half space occupied by fore-central eyes; pars thoracica oval, convex; fovea oval, longitudinal; radial and caput striæ fairly well defined; profile-contour represents an angle of 45° at posterior inclination, occiput somewhat horizontal, perceptible double curve, slopes across eye-area.

9

Eyes pearl-grey, on black rings; posterior row moderately procurved, centrals on oval spots; of about equal size, nearly equidistant; space between median pair, which are visibly the most distant apart, less than an eye's breadth; anterior row slightly recurved, centrals more than half size of hind-pair, separated from each other and side-eyes by an interval equal to two-thirds their diameter; laterals posited obliquely on separate low tubercles, fully the radius of a fore-eye apart: latter eyes suboval, visibly larger than hind-pair.

Legs fulvous, blackish annulations, more or less faint; annuli on coxal joints; rings evanescent on two first pairs; moderately slender; hairs fine, sparse; spines slender, black; about 6 or 7 on femora + tibiæ; single spine on patellary joints; about 7 on metatarsi of two first pairs; 6 or 7 and ring of 4 on hind metatarsi.

Palpi, colour and armature of legs; slender, length 9mm., equal to metatarsus + tarsus of a fore-leg; pars humeralis more than one-third longer than cubital + radial joints; pars cubitalis subquadrate, more than half length of penultimate joint, projects black bristles; radial joint moderately incrassated forwards, prolonged on outer side into a subquadrate, black-margined, membranous process, about twice as long as broad, apex grooved; immediately beneath is a stout darkish-brown process; posited on superior angle of article is a strong, acute, downward-curved, blackish process; below and contiguous to the latter is an oblong dark projection, whose short, pointed angles are connected by a U-shaped costa on its outer face; digital joint about twice as long as the pars radialis, bulb one-fourth length of article; genital bulb orange-brown, fading into a yellowish stone-colour towards fore-end; base semi-globose, anterior half rather more, depressed, displays round margin of lamina a wide, everted, brown border; projecting forwards from centre of bulbus is a large, membranous, sinuating process, base colour of bulb, fore-end blackish-lake; flanked by two acute, somewhat ear-shaped apophyses, reaching forwards to about one-half its own length; outer apophysis yellowish, apex brownish, inner green tinge, passing into dark-green on second half: lamina ovate, above bulbous, strongly convex, prolonged in a cylindrical form for 2·5mm., extension resembles digital joint of female's palpus; armature fine hairs, long slender bristles, three strongish spines at extremity; reddish-brown, slender extremity fulvous.

Falces brownish-orange; project forwards at an angle of 30°, divergent, of somewhat even breadth, second half curved, extremities—which are dilated on inner side—directed towards each other; small oval, plano-convex protuberance beneath angle of caput: fangs long; falx equal to cephalothorax in length, armed with only 2 short teeth projecting from beneath fore-end, first tooth longest.

Maxillæ fulvous, long, moderately enlarged forwards, obtusely pointed, somewhat curved towards each other.

Labium deeper tone, clouded ; length somewhat surpasses width, subquadrate, emarginate, nearly one-half length of maxillæ.

Sternum fulvous, margins clouded ; broad-cordate, eminences opposite coxæ.

Abdomen elongate-oviform, lateral margins and posterior third rugose ; light yellowish-brown, shading off to black-brown at posterior end, dappled with pale flecks.

Femina.—Ceph.-th., long, 5 ; broad, 3.   Abd., long, 6 ; broad, 3.   Legs, 1, 4, 2–3 = 23·5, 20, 16 mm.

Cephalothorax fulvous, median band shaded with dark-brown, bifurcates at the red thoracic indentation, rapidly compressed at posterior centre eyes ; marginal band broad, similar shade ; almost glabrous; length equals tibia of a fourth leg ; cephalic region quadrate, lateral index equals breadth of hind-row of eyes ; height of *clypeus* exceeds diameter of a fore-central eye ; thoracic part oval, indentation narrow, longitudinal ; radial striæ very perceptibly stronger than caput grooves ; profile-line slopes slightly from hind-row of eyes, dips posteriorly at an angle of about 40°.

Eyes grey, on blackish oval spots, enclose an oval space ; posterior row of moderate and about equal size, equidistant, an eye's breadth apart ; centrals of anterior row smaller than hind-pair, separated from each other by a space equal to three-fourths their breadth, and from side-eyes by rather more than an eye's diameter ; laterals posited on low blackish tubercles, divided by an interval equalling two-thirds breadth of the hind-eye ; fore-eye oval, rather the largest of eight.

Legs in colour and armature do not differ essentially from male's ; second leg slightly exceeds the third in length ; superior tarsal claws—first pair strong, moderately curved, 12 teeth increasing in length and strength ; inferior claw stout, sharply bent, 3 strong backward-curved teeth.

Palpi fulvous, annulations evanescent ; few spines on all joints ; palpal claw strong, 8 comb-teeth increasing in length, directed forwards ; free end moderately curved, nearly one-half length of claw.

Falces yellowish-brown ; black hairs ; inclined forwards and outwards, conical ; first half of profile arched ; short, tapering (plano) process beneath angle of caput.

Maxillæ yellow-brown ; black hairs ; dilated forwards, obtusely pointed, inclined towards each other.

Labium greenish tinge ; oval, strongly emarginate.

Sternum fulvous, bordered by a deeper shade; cordate, nearly as wide as long, well-developed eminences opposite coxæ.

Abdomen narrow, oviform; second half and lateral margins rugose; light yellow-brown, shading off to black-brown on posterior third; base dappled with a lighter brown; lightish-yellow parts spotted with dark-brown; ventral shield yellow-brown, fuscous spots; hairs sparse, golden. *Corpus vulvæ* transverse, prominent, reddish-brown, oval elevation, bears on face a somewhat dagger-shaped depression, hilt and quillon represented by three oval foveæ, blade tapers to a point above the rima genitalis.

This species frequents the loose bark of *Fuchsia excorticata;* the female fabricates a globose cocoon 10mm. in diameter, of a hard, parchmenty texture, approximating in colour to its surroundings, suspended by a short web; male examples were generally captured with the females (January). The somewhat extensive sheet of web is composed of a fine open mesh, and attached to the surrounding objects by fine lines. Captured in the forest on Mount Egmont, *A. T. U.*

Fam. ENYOIDÆ.

Gen. Habronestes, L. Koch.

Habronestes celeripes, sp. nov.   Plate XXI., fig. 2.

Mas.—Ceph.-th., long, 2·1; broad, 1·7.   Abd., long, 2·6; broad, 1·5.   Legs, 1, 4, 2, 3 = 10, 7·8, 7·2, 5·5 mm.

Cephalothorax pale-brownish flesh-colour, cephalic part stained with brownish-yellow; from limit of caput two brown lines project forwards, curving upwards midway to posterior row of eyes; streaks of a lighter tone nearly connect these lines with side-eyes; lightly-shaded semi-oval patch encloses the red thoracic indentation; thorax displays two bands, marginal faint, submarginal resolved into a few dark spots; hairs white, compound-sessile, extend along unshaded ridge of caput; pars cephalica strongly convex, squarely truncated, facial index perceptibly shorter than lateral; *clypeus* nearly vertical, depth rather surpasses space occupied by fore-centre eyes; pars thoracica conoid, rises abruptly from margin, apex at limit of cephalic part; narrow longitudinal groove represents thoracic indentation; radial striæ well defined; profile-contour rises at an angle of 55°, visibly inclined forwards across second half of caput, fore-end curved.

Eyes on blackish spots; hind-centrals oval; posterior row rather strongly procurved, fore-margin of centrals being in line with hind-margin of laterals; of moderate and nearly equal size; median pair separated by rather more than an eye's breadth, divided from laterals and fore-centrals by intervals equalling their diameter and a half; anterior row pro-curved, fore-margin of centrals in line with space between laterals; median pair visibly less than an eye's breadth from one another, rather more than their diameter from side-eyes; laterals nearly equal centrals in size, posited obliquely on very low eminences, separated by a space equal to two-thirds breadth of a fore-lateral eye.

Legs shade yellower than cephalothorax; annuli some-what obliterated, orange-yellow, deepening in tone forwards, shaded on posterior pairs; two central annulations on femora; three rings on tibiæ + metatarsi; moderately slender; hairs fine, outstanding; 7 or 8 short spines on femoral joints; patellæ, 2; tibial joints, about 7; metatarsi show 5 or more irregular spines, ring of 4 at extremity; superior tarsal claws —first pair rather slender, curved, 13 long, somewhat even, open comb-teeth, exceeding length of claw; inferior claw shortly and sharply bent, 2 points.

Palpi fulvous, except radial + digital joints, which have a deeper brownish tinge; humeral joint strong, linear, fully equal in length to cubital + radial together; projects 3 black bristles in line on superior surface; pars cubitalis somewhat slighter, 2 bristles; pars radialis rather surpasses former article in length, about as broad as long near base; bifurcates, pro-longed outwardly into a large, pointed, ear-shaped process, posterior and upper margin armed with a series of black, acute, tooth-like projections, basal tooth much the longest and stoutest; projecting from base of process, lower side, is a brown, translucent, stout, outward-curved apophysis; some-what slender at articulation with digital joint; few bristles; pars digitalis scarcely exceeds two former articles in length; lamina ovate, base slender, tapers somewhat sharply at ex-tremity; moderately clothed with fine hairs; genital bulb brownish; viewed partially from front, shell-shaped, wide and convex posteriorly, concave within, curved forwards, upper margin tapers beneath lamina; lower margin pinched into an angular form; bulbus produced towards fore-end, beneath, into a strong, blunt, forward-curved process of its own colour; projecting from within fore-half of bulb are, apparently, two membranous lobes; divergent extremities of first or outer lobe, dark, recurved; second lobe pale within, broad, some-what triangular on inner side, displays at apex a short, pointed, black process; a long, fine, black apophysis curves across face of genital bulb, close to margin of lamina, from inner side.

Falces red-mahogany colour; linear-conical, vertical; basal third, superior side, displays a plano-conical elevation; as stout as thigh of a fore-leg, length equals humeral + cubital joints of palpus; fangs short, slight.

Maxillæ yellow-brown, fore-third brownish; about twice as long as wide, straight on inner side, curved outwardly; inclined towards each other.

Labium colour of maxillæ, base darkest; rather longer than broad, oval, apex concave.

Sternum cordate.

Abdomen slender, oviform, tapers to base, convex above; stone-green; basal fourth of folium ovate, posterior three-fourths lanceolate, margins acute-crenate; oval part spotted with black dots; lanceolate extremity shows large, pale stone-coloured flecks; border of folium defined by greenish-black, confluent, mottled blotches; lateral margins lightly speckled with green-black; a shaded undulating band extends from base to spinners; specific pattern more or less picked out with white, compound-sessile hairs.

Femina.—Ceph.-th., long, 2·4; broad, 1·6. Abd., long, 2·5; broad, 2. Legs, 1, 4, 2, 3 = 6, 5·7, 5·2, 5 mm.

Cephalothorax pale-brownish flesh-colour, cephalic region partially suffused with a brownish-yellow, brown lines curve forwards above lateral grooves from posterior extremity of caput; similar streaks on fore-margin; lightly-shaded semi-oval mark encloses red indentation on thorax; marginal band pencilled, submarginal band resolved into brown spots; few black bristle-like hairs and white compound-sessile hairs on caput, latter extend along unshaded median line; pars cephalica strongly convex, squarely truncated, lateral index fully equals facial; *clypeus* vertical, in height rather shorter than interval between fore- and hind-median eyes; pars thoracica oval, conoid, indentation longitudinal, grooved; contour of profile rises from thoracic junction at an angle of 55°; second half of cephalic part straight, perceptibly inclined forwards, fore-end curved.

Eyes in size and position resemble male's.

Legs yellowish flesh-colour, annulations more or less effaced, orange-yellow, darker and more pronounced on metatarsal joints; spine armature does not differ essentially from male's, also form and pectination of claws.

Palpi yellowish, fore-half of pars digitalis brown-lake; somewhat sparsely armed with hairs and bristle-like spines; moderately slender, length equals metatarsus + tarsus of first leg.

Falces red-mahogany colour; conical, vertical, tumid at base in front; plano-convex conical protuberances on sides.

Maxillæ brownish-yellow, pale greenish-yellow apices; taper perceptibly to base, rounded on superior side, nearly twice as long as broad, inclined towards each other.

Labium brownish-lake, apex greenish : oval, rather longer than wide, apex concave.

Sternum fulvous ; cordate, width between coxæ of first pair rather exceeds one-half its length.

Abdomen oviform, basal extremity broadest, convex above ; ground-colour green ; anterior fourth of folium ovate, posterior three-fourths broad-lanceolate, acute-crenate ; margin defined by greenish-black blotches and spots, more or less effaced towards spinners ; lateral margins spotted, normal coloration, series of confluent spots form an undulating longitudinal band ; ventral surface light greenish-brown ; fairly clothed with brownish hairs, pattern partially picked out with white, compound-sessile hairs. ‑ *Vulva* brownish-yellow, glossy ; scapus springs from a slight elevation, large, tumid, about one-third wider at base than long, rather narrower and rounded in front, curves towards the rima genitalis ; the somewhat cylindroid lateral margins terminate in subfree, ovate, flatly-convex apices, directed towards each other.

This pretty little species was not uncommon amongst vegetation growing over fallen trees.

Mount Egmont, Stratford, *A. T. U.*

Habronestes scitula, sp. nov. Plate XXI., fig. 5.

Femina.—Ceph.-th., long, 1·9 ; wide, 1·2. Abd., long, 2·2 ; wide, 1·5. Legs, 4, 1–2–3 = 5·5, 4·9 mm.

Cephalothorax brownish-yellow, dorsal aspect of cephalic region coffee-brown, displays a narrow dark medial streak ; dorsal band streaked mahogany-brown, broad, extends from lateral eyes to base of thorax, intersected by a tolerably wide line of the normal ground-colour, which bifurcates forwards from limit of caput ; marginal zone dark-brown, narrow ; cephalic area clothed with few whitish hairs, chiefly on the lighter parts ; ovate, lateral compression at caput slight ; pars cephalica moderately convex, sides abrupt, roundly truncated ; *clypeus* squarely truncated, directed outwards, height nearly equal to diameter of a lateral eye ; thoracic part slopes somewhat abruptly from summit, posterior incline scarcely steeper than lateral ; slight longitudinal reddish groove on crown of thorax ; striæ well defined ; profile-contour rises at an angle of 60° from stalk, slopes moderately across occiput, eye-region rounded.

Posterior centre *eyes* and laterals large, of about equal size, nearly equidistant, enclose a subcircular space, perceptibly wider than long ; anterior centrals very small, separated from each other by nearly an eye's breadth, somewhat less than that interval from side-eyes of same row, which are divided by a space exceeding their diameter ; line drawn across fore-margin of anterior centrals intersects fore-laterals.

Falces brownish-orange, oblong-oval, olive-brown mark on face; conical, directed inwards, stout, occupy breadth of clypeus; in length nearly equal radial + digital joints of palpus.

Maxillæ fulvous, olive-green tinge, base clouded with dark-brown; rather longer than broad, dilated somewhat forwards, obtusely pointed, inclined towards each other.

Labium similar coloration; conical, abscinded, length somewhat surpassed by breadth, rather more than one-half length of maxillæ.

Sternum yellowish, margin dark-chocolate, suberenate; two somewhat angular interrupted brown bars in centre; two dots near the sharply-pointed apex; broad-cordate.

Legs greenish-yellow, spots and annulations olive-brown; femora have basal and nearly central annuli, resolved more or less into spots on superior aspect; tibial joints, ring on first half, somewhat broken into spots on first and second pairs; metatarsi show central and distal annulations of a lighter shade; hairs whitish, sparse; spines fairly numerous on all joints except tarsal; spines long and strong, metatarsal longest. Superior tarsal claws—fourth pair, rather fine, evenly curved, 5 open teeth; inferior claw fine, sharply bent, teeth (?)..

Palpi colour of legs; hairs sparse; few bristles; moderately strong, less than twice length of falces; palpal claw, well curved, 2 long curved teeth.

Abdomen ovate; greenish stone-colour, passing into a darker tone on lateral margins; anterior fourth stained with lake; markings dark olive-brown; dorsal aspect spotted, sparingly in centre, more thickly at extremities; basal end bordered by bars increasing in width forwards; lateral margins marked with interrupted, irregular, oblique streaks, terminating on posterior half in four well-defined, spotted, lanceolate figures encroaching on dorsum, inclined forwards; ventral field similar hue to dorsal, few spots; lighter parts of abdomen, except central area, moderately clothed with white hairs. *Vulra* represents a large, transverse, oval, lip-like projection, depressed and yellowish in centre; lateral margins reddish-brown, tumid.

Single example, taken in the forest near Stratford, *A. T. U.*

Fam. THURIDIIDÆ.

Gen. ARIAMNES, Th.

Ariamnes flavo-notatus, sp. nov.

Femina.—Ceph.-th., long, 1. Abd., long, 1·6; depth at spinners, 1·1. Legs, 1, 2, 4, 3 = 8, 4·6, 3·4, 2·3 mm.

Cephalothorax brown, suffused with blackish-purple: transversely rugulose; few bristles on caput; broad-ovate, lateral constriction at caput slight; pars cephalica convex, ocular eminence rather prominent; *clypeus* visibly inclined forwards, height exceeds one-half depth of eye-area ; pars thoracica convex, fovea oval, small ; normal grooves moderately prominent; contour of profile represents a strong curve.

Eyes rather large ; fore-pair dark ; posterior row slightly procurved ; centrals perceptibly smaller than laterals, separated from each other by a space scarcely equalling an eye's breadth, about twice that distance from side-eyes and fore-centrals, latter interval somewhat the shortest; anterior row recurved, median pair about one-third smaller than hind-pair, one-fourth their diameter from each other, separated from side-eyes by two-thirds their space ; laterals posited obliquely on a stony tubercular eminence, contiguous.

Falces yellowish mahogany-colour ; transversely rugulose : broad, somewhat flat, gibbous at base in front, vertical, equal digital joint of palpus in length ; breadth more than one-half length, second half rounded on inner side, directed somewhat outwards, 4 stout teeth in outer row.

Maxillæ fulvous ; somewhat linear, fore-half broadest and sharply bent over lip.

Labium, base fuscous, margins brownish ; wider than long, tumid and somewhat truncated, transverse groove.

Sternum chocolate-brown ; broad-cordate, studded with papillæ.

Legs brownish-yellow, three faint annuli on femora ; patellæ brown ; tibial joints have central and apical rings, narrower and more pronounced on hind-pairs ; long, slender ; armature dark hairs, few slender erect bristles.

Palpi straw-colour, except digital joints, which have a mahogany shade; slender, pars digitalis equals cubital + radial joints in length ; few long hairs and fine bristles.

Abdomen, fore-half dark-brown, passing into a pale shade on posterior half and lateral margins, spotted with large stone-coloured flecks ; elongate-ovate, profile somewhat triangular ; distance from the lower angle—from which the rather long spinners project—to the rounded posterior extremity is slightly shorter than dorsal line. *Corpus vulvæ* yellowish ; represents a transverse oval area, occupied by large ovate foveæ, intersected by a rather broad septum ; a reddish-brown bead-like pimple projects above base of septum.

Single specimen captured in the forest near Stratford, *A. T. U.*

Gen. LINYPHIA, Latr.

Linyphia sennio, sp. nov.   Plate XXI., figs. 15, 16.

Mas.—Ceph.-th., long, 2·1 ; broad, 1·8. Abd., long, 2·2 : broad, 2. Legs, 1, 2, 4, 3 = 13, 10, 7, 5·5 mm.

Cephalothorax brownish-orange, light-fuscous shading mostly about median line, normal grooves and marginal zone; line composed of lake-coloured dashes bifurcates at limit of caput, joins posterior lateral eyes; triangular lake mark beneath fore-centrals; few black bristles, chiefly about occiput; oval, lateral margins of caput moderately compressed; pars cephalica convex, roundly truncated, eye-eminence prominent, angles cup-shape; facial index perceptibly exceeds lateral; *clypeus* inclined forwards, sides less tumid than female's, height equal to two-thirds space of fore-centre eyes; pars thoracica convex, fovea oval, deep, longitudinal; caput and radial striæ faint; contour of profile slopes across caput, perceptibly curved; angle of posterior inclination about 45°.

Posterior row of *eyes* slightly procurved, median pair scarcely their diameter from each other, three-fourths their space from lateral eyes; anterior row recurved; centrals dark, largest of eight, separated from one another and side-eyes by an interval slightly exceeding their own breadth; laterals contiguous, seated obliquely on a common. fair-sized, lake-coloured tubercular prominence.

Falces red-mahogany, passing into a darker shade at extremity; transversely rugulose; linear, vertical, moderately stout, length nearly equal to the pars humeralis of palpus.

Maxillæ long, linear-oval, somewhat pointed, inclined towards each other.

Labium nearly as long as broad, roundly pointed, everted, less than one-half length of maxillæ; organs orange-brown. base suffused with lake.

Sternum lake-colour, centre band yellowish, cordate, eminences opposite coxæ.

Legs light orange-yellow, lake stains on coxal joints; annuli light cinereous-brown, bordered and more or less suffused with lake; rings on thighs more or less evanescent, resolved into spots; two on patellar joints; four annulations on tibiæ of fore-pairs, three on hind-pairs; metatarsal joints have two annuli; hairs somewhat sparse, outstanding; spines long, slender, mostly project laterally; 4 spines on femora; patellæ, 2; tibiæ of first pair, 15; second, 10; metatarsi of first, 13; second pair. 8; tibial joints of hind-pairs, 5; metatarsi, 2 spines.

Palpi shade paler than legs; slender; pars humeralis rather surpasses cubital + radial joints together in length, thickens slightly at fore-end; cubital joint tapers somewhat to base, more than one-half length of penultimate article, projects two bristles; radial joint lageniform, fore-half tumid, prolonged above outer side into a reddish, flat, tapering process;. base slender; projects above long bristles; digital joint subglobose, about length of pars radialis; lobes lake-brown, appendages lake-black ; laminæ bulbi yellowish-brown, passing into olive-green on fore-end ; directed towards each other ; hairs coarse, sparse ; inversely ovate, broad, strongly convex ; rise above bulbus ; lamina on outer side exhibits a large moderately-deep concavity—segment of a circle—of nearly its own length, margin—base of segment intact, lake-colour ; base of lamina at termination of cavity prolonged into a twin, submembranous, brownish, curved process, apices directed outwards ; projecting from a ring on summit of lamina is a long, conspicuous, dark, cylindrical process, obliquely truncated, on its outer side, for nearly whole length ; lobes of bulb wide, subvertical ; first lobe of somewhat even width, figured with a blackish ciliate-like mark ; second lobe ·rapidly compressed below, prolonged into a strong, black, forward-curved apophysis ; above the latter organ is a large, broad, pitted, brownish-yellow, up-curved process ; anterior appendage, viewed somewhat in front, represents a rather large compressed or pinched membranous process curving downwards to the large yellowish process.

Abdomen broad-oviform, depressed, sides abrupt, humeral tubercles large, conical, project outwards and backwards ; profile of abdomen somewhat diamond-shaped ; sparingly clothed with orange and black bristle-like hairs ; stone-colour, approximating to olive-green : design formed by a series of re-curved irregular bands, composed of creamy-stone, purple margined and spotted, free and coalescing flecks ; bands angular and more defined on posterior slope ; lateral margins bordered by a brown band without very determinate limits ; ventral field olive-stone colour, displays a broad transverse band of mixed normal colours.

Femina.—Ceph.-th., long. 3 : wide, 2. Abd., long. 3 : wide, 2·5. Legs, 1, 2, 4, 3 = 14. 10·1, 8·8, 6·5 mm.

Cephalothorax brownish-orange, lightly shaded, chiefly on grooves and margins, with black-brown ; splashed, lake-coloured, bifurcating line extends from limit of caput to posterior lateral eyes ; triangular lake mark on *clypeus ;* hairs sparse, black, bristle-like ; cephalothorax equal in length to metatarsal joint of first leg ; oval, fairly compressed forwards ; pars cephalica strongly convex, roundly truncated, lateral index scarcely equals facial ; eye-eminence well developed ; *clypeus* outwardly inclined, depth exceeds space between a fore-central and lateral eye next to it ; an angular furrow beneath anterior row of eyes gives its tumid sides a conical (plano) form ; pars thoracica convex ; large circular depression on posterior slope ; normal grooves faint ; profile-contour rises from thoracic junction at an angle of 45°, slightly curved across caput.

Posterior row of *eyes* perceptibly procurved, centrals separated by an interval equal to an eye's breadth, their space from laterals; anterior row recurved, median pair larger than posterior centrals, somewhat less than their diameter from each other, rather more than that distance from hind-pair, interval between them and side-eyes surpasses one-half their space; laterals have the opalescent lustre of posterior centrals: smallest of eight, posited obliquely on a low, common tubercle. contiguous.

Falces red-mahogany, deepening in tone at fore-end: moderately strong, vertical.

Maxillæ light brownish-yellow, suffused with lake-brown: linear-conical, inclined moderately over *labium*, which is oval, about half as long as maxillæ, similar colour.

Sternum brown-yellow, suffused with lake; cordate, eminences opposite coxæ.

Legs yellow-orange, coxæ blotched with lake; femora have four interrupted light cinereous-brown rings, bordered with lake; tibiæ of fore- and hind-pairs have respectively 3 and 4 annulations; metatarsi, 2; hairs somewhat sparse; thighs armed with few black spines; patellæ, 2; tibiæ of first and second, 15 + 8, hind-tibiæ 4 + 8, long black spines; metatarsi of first pairs, inner row of 6 projecting spines surpassing tibials in length, 6 outer somewhat shorter; metatarsal joints of second pairs, 4 + 6 similar spines; metatarsi of hind-pairs have 2 centrally-placed strongish spines. Superior tarsal claws —first pair, somewhat shortly bent, 4 teeth increasing in length and strength; inferior claw long, fine, sharply bent, small points.

Palpi coloration of legs; strongish bristles; slender, as long as cephalothorax; palpal claw less bent than tarsal; 4 teeth similar in form to teeth of tarsal claws.

Abdomen does not differ essentially in form or coloration from male's. *Corpus vulvæ* orange-brown, clouded with brownish-lake; represents a large transverse oval eminence: somewhat clathrate; median line concave; above the rima genitalis is a large, rather deep, oval fovea, laterally bounded by blackish tumid costæ; a semi-free dark costa forms a somewhat depressed arch in line with margin of fovea, the involute basal ends confluent with outer costæ.

Several examples of this brightly-coloured species were captured amongst shrubs in the forest near Stratford, and on Mount Egmont, *A. T. U.*

Linyphia multicolor, sp. nov.

Mas.—Ceph.-th., long, 1·5; broad, 1·1. Abd., long, 1·7: broad, 1·3. Legs, 1, 4, 2, 3 = 10, 6, 5·5, 4·5 mm.

Cephalothorax brownish amber-colour, streaked with lake, lateral borders lightly suffused with a blackish-olive; median band tapering, stretches from frontal margin to fovea, mottled with a deeper hue; glabrous; broad-ovate, depressedly convex; ocular eminence prominent; *clypeus* nearly vertical, height equals one-half facial space; thoracic fovea oval, deep; caput and radial striæ tolerably strong; profile-contour represents a slight posterior incline.

Eyes rather large, hind-centrals and laterals opalescent : posterior row slightly procurved, median pair about one-fourth larger than fore-centrals, removed from laterals by an interval scarcely equalling their diameter, and from each other by rather more than one-fourth that space ; anterior row recurved, centrals posited somewhat obliquely on strong prominences, nearly twice·their breadth apart, separated by rather more than that distance from hind-pair, about their radius from side-eyes ; laterals fully as large as anterior centrals, seated obliquely on a strong eminence, contiguous.

Falces light-red mahogany: of somewhat even width, moderately gibbous at base in front, second half divergent, perceptibly inclined outwards. moderately stout, length equals twice depth of clypeus.

Maxillæ yellowish-mahogany. suffused with olive on first half; spathulate, base dilated, long. moderately inclined towards each other.

Labium fulvous. nearly twice as wide as long, rounded, perceptibly emarginate, scarcely reaches second half of maxillæ.

Sternum brownish, fulvous. heart-shaped figure with seven acute projections occupies centre: broad-cordate.

Legs and cephalothorax concolorous, annuli light chocolate-brown ; central and distal rings on femoral joints tinged with olive-green ; three annulations on tibiæ, basal more or less obliterated ; centre and apical rings on metatarsi ; hairs sparse ; very few bristle-like spines, longest and strongest on patellæ, which project two.

Palpi colour of legs ; humeral joint about one-third longer than two following articles: pars cubitalis dilated forwards, projects a long bristle ; pars radialis rather surpasses former article in length, cup-shaped, deeply emarginate at insertion of lamina, few long bristles ; laminæ bulbi orange-brown, hairs fine, sparse ; directed towards each other ; basal two-thirds suboborate, broad, fore-third moderately compressed, tumid, upturned ; genital bulb colour of lamina, cap consists of two membranous lobes, base of lowest produced into a short, dark, triangular process ; suspended from front of bulb is a wide, sharply-constricted, dark-margined, membranous process, directed downwards and outwards ; next to it is a larger appendage of somewhat similar form, whose compressed ex-

tremity is wider and more distinctly curved; base enclosed by
a subtriangular dark-bordered lobe.

Abdomen oviform, strongly convex, almost glabrous, folium
extends over first half, sublanceolate, olive-brown, series of
free and coalescing fuscous flecks along margins; the blackish
petiole exhibits on either side a conspicuous oblique white bar;
second half of dorsum displays an olive-brown, dark-spotted,
oval mark; median band creamy-white, lake stains, fore-half
formed by two triangular figures enclosing two lake-coloured
dots; lateral margins light olive-brown, dark fuscous spots interrupted by creamy marks, most conspicuous along border of
dorsal pattern, lake stains; blackish olive-green shield on ventral surface, constricted above spinners, white spot in each
indentation.

Single example, captured in the forest near Stratford,
A. T. U.

Linyphia cruentum, sp. nov.

Mas.—Ceph.-th, long, 2; wide, 1·8. Abd., long, 2·3; wide,
1·2. Legs, 1, 2, 4, 3 = 12·5, 7, 6·5, 5 mm.

Cephalothorax fulvous, splashed with lake, median band
shaded with brown, tapers slightly from hind-lateral eyes to
base; few coarse hairs on caput; clathrate; pars cephalica
convex, ocular prominence moderate; facial index equals one-
half breadth of thorax, lateral index under two-thirds of former;
clypeus convex, depth equal to two-thirds length of eye-area;
thoracic part broad-oval, fovea transverse broad-oval; normal
grooves well defined; profile-contour slopes backwards with a
slight double arch.

Eyes of moderate size, on dark rings; posterior row perceptibly procurved; centrals rather less than an eye's breadth
apart, a diameter and a quarter from laterals, visibly less than
that interval from fore-centrals; anterior row moderately recurved, median pair rather smallest of eight, separated by
rather more than their diameter, three-fourths their breadth
from side-eyes; laterals have the pearly lustre of hind-
centrals, posited obliquely on a dark common tubercular
eminence, contiguous.

Falces colour of cephalothorax; outwardly inclined, somewhat gibbous at base, second half tapering, curves upwards
and outwards.

Maxillæ fulvous, yellowish reflections, passing into mahogany-brown at apices; long, somewhat enlarged at extremities, pointed, rounded on superior side, sharply truncated on
inner side, inclined towards each other; furnished with stiff
bristles.

Labium and maxillæ concolorous; more than one-third

length of latter, large, rather wider than long, roundly emarginate.

Sternum light olive-brown, spotted and bordered with brown; subcordate, nearly as wide as long between first pair of coxal joints.

Legs orange-yellow, passing into orange-lake on fore-ends of femoral joints, first pair most shaded; tibiæ have two or three not - well-defined orange-lake annulations; metatarsi two; legs strong; bristle-like hairs on femoral joints, more especially clustered on inner side of first pair, and on superior surface of hind-pairs, which are armed beneath with a double row of bristle-like spines, strongest on second and third legs; patellæ, 1 spine; single row beneath tibiæ + metatarsi.   Claws weak.

Palpi light orange-yellow; pars humeralis fully one-third longer than cubital + radial joints; pars cubitalis short, dilated forwards, projects a strong black bristle; pars radialis twice length of former article, outer side prolonged, much enlarged forwards, rounded, armed at extremity with a row of strong long bristles; viewed from inner side somewhat cup-shaped, short; digital joint about as long as cubital + radial joints together, well developed; laminæ bulbi reddish mahogany-colour; moderately furnished with stiff black hairs; oval, directed towards each other; genital organs partially covered by a brownish-yellow, spotted membrane, projecting forwards from beneath is a greenish, elongate, pointed process, which, viewed from beneath, discloses a membranous truncated process, margins involute; immediately under it is a brownish curved lobe, convex side directed backwards; extremities prolonged, free, upper tapering, lower—not visible from outside—somewhat membranous; bulb partially enclosed on lower side by a greenish, somewhat oval, or pointed membrane.

Abdomen elongate-oviform, yellowish stone-colour; folium covers dorsal region, deeper tone, green tinge, second half stained with lake, margins sinuating, olive-black, interrupted, centre pair of curves creamy-white; median band exhibits a series of six large creamy-white spots on first quarter; irregular-shaped figure near centre, of similar colour; second half displays three tapering groups of coalescent, yellow, lake-tinged spots; oblique brown marks on lateral margins; ventral shield brown.

Forest, Stratford, *A. T. U.*

Linyphia albi-apiata, sp. nov.

Mas.—Ceph.-th., long, 1·7; wide, 1·1.   Abd., long, 1·5; wide, 1.   Legs, 1, 4, 2, 3; 1st pair, 5·4 mm.

Cephalothorax light-brown, green tinge, shaded, especially on lateral margins and radii, with olive-brown : clathrate : oval ; pars cephalica convex, not constricted at junction with thorax, lateral index surpasses facial, ocular eminence prominent ; *clypeus* subvertical, depth equal to two-thirds length of eye-area ; pars thoracica convex, fovea deep, somewhat triangular, apex directed forwards ; caput striæ strong, thoracic moderate ; profile-line rises moderately, with an even contour, from the stalk, rounded across occiput.

Eyes on dark rings, enclose a linear-oval space ; posterior row of tolerable and about equal size ; centrals scarcely two-thirds of an eye's breadth apart, one-fourth more than their diameter from laterals ; median pair of anterior row dark, much the smallest of eight, separated from one another by a distance perceptibly less than their radius, and from side-eyes by an interval surpassing their diameter, divided by more than their space from hind-centrals ; laterals have the pearl-grey lustre of posterior median pair, fore-eye somewhat the largest, seated obliquely, about one-fourth their breadth apart, on a low, dark, tubercular eminence.

Falces brownish, lightly clouded ; rugulose : strongly inwardly inclined ; base tumid, fore-third divergent ; about twice as long as broad at base ; length exceeds width of eye-area by nearly one-third.

Maxillæ yellowish-brown ; stout, fore-half dilated, pointed ; somewhat inclined towards each other.

Labium fuscous ; pale apex ; perceptibly broader than long, margin tumid, everted.

Sternum dark mahogany-brown ; cordate.

Legs yellowish, annulations greenish-black ; femora show basal and subcentral rings, faint on two first pairs ; tibiæ + metatarsi have two rings, first on basal half, second at apex ; tarsi one : legs moderately stout, first and second somewhat exceed third and fourth in length and strength ; hairs sparse, fine ; few erect, black, bristle-like spines on all joints except tarsal.

Palpi, humeral joint yellowish, surpasses cubital + radial together in length ; cubital joint broad : radial pale-straw-colour ; few bristle-like hairs ; wide at extremity ; pars digitalis about one-third longer than penultimate article ; lamina oval, tapering, prolonged for nearly half its own length beyond bulbus, armed with strong bristles ; accessory lamina oval, projects somewhat beneath, extends nearly half-way across bulb ; genital bulb yellow, fore-end and apophysis splashed with red : subobovate, depressed above, prolonged into a rather wide, grooved, subfree apophysis, which is sharply bent backwards on superior face nearly to base of bulbus, from thence bent upwards, its short darkish free end curving over its own base, which projects a rather short, black, curved process from beneath the free end.

Abdomen oviform ; dorsal field light stone-colour, flecks lobate, large, creamy-white, more pronounced round foremargin ; displays a black sublanceolate figure in front, haft terminates half-way to spinners; black spot on either side close to lance-head ; sides bordered by two wide bands, upper greenish-black ; lower brownish, white flecks ; ventral surface brown.

Femina.—Ceph.-th., long. 1·8; broad, 1. Abd., long. 1·6 broad, 1. Legs, 1, 2, 4, 3. Leg of 1st pair, 4·5 mm.

Cephalothorax reddish-brown, shaded, chiefly on margins and grooves, with a darker tone ; almost glabrous ; clathrate ; oval, constriction at caput slight ; pars cephalica strongly convex, truncated ; lateral index about equals facial ; *clypeus* directed forwards, slightly shorter than depth of ocular area ; pars thoracica convex, fovea elliptical, longitudinal ; normal grooves fairly strong; profile-contour rises at an angle of 40°, almost horizontal across occiput, slopes at eye-region.

Eyes rather large ; posterior row slightly procurved, median pair an eye's breadth from laterals, about two-thirds that space from each other ; anterior row sensibly recurved, centrals about one-third smaller than posterior pair ; scarcely their radius from one another, removed from side-eyes by an interval equal to their diameter, and from hind-centrals by rather more than that space ; laterals have the pearl-grey lustre of posterior median eyes, surpass them in size by about one-third, placed obliquely, one-fourth their breadth apart, on moderately stout tubercular prominences.

Falces orange-brown; clathrate; tumid at base, taper somewhat, divergent, vertical, stouter than the femur of first leg, nearly as long as radial + digital joints of palpus.

Maxillæ yellow-brown, clouded with brown : strong, forehalf dilated, pointed, moderately inclined towards each other.

Labium, base dark-brown, margins yellow-brown ; rather wider than long, margin tumid, everted.

Sternum dark mahogany-brown ; clathrate, cordate, conoid prolongation between posterior coxæ.

Legs brownish-yellow, moderately wide greenish-black annulations ; basal and nearly central rings on femora, faint on two fore-pairs ; tibiæ + metatarsi have central and apical annuli ; legs strong, first and second, third and fourth do not differ much in length or strength ; hairs rather sparse; femoral spines short, sparse ; genual, tibial, and metatarsal spines tolerably long and numerous.

Palpi colour and armature of legs ; slender, about equal to cephalothorax in length.

10

Abdomen oviform ; dorsal area light stone-colour, flecks
rather large, lobate, creamy-white ; somewhat ill-defined black
lanceolate figure on fore-part, haft extends midway to spinners ;
black spot on either side near base of lance-head ; lateral
borders display two broad bands, superior band greenish-black,
inferior stone-colour ; ventral surface brown ; *corpus vulvæ*
pale-brown, mottled with greenish-black ; transverse oval area,
projects over the rima genitalis, indented by two yellowish
foveæ ; lateral margins shortly involute, red-tipped ; centrally
produced into a yellowish, moderately broad and long, emarginate, upcurved scapus.
    Stratford, *A. T. U.*

Linyphia pellos, sp. nov. Plate XXI., fig. 10.

Mas.—Ceph.-th., long, 2·5 ; wide, 1·8. Abd., long, 2·9 ;
wide, 1·8. Legs, 1, 2-4, 3 = 9·8, 7·5, 6·8 mm.

Cephalothorax creamy stone-colour, base of caput displays
oval patches of a gamboge hue, with rather faint blackish
margins on outer side, prolonged forwards into angular marks ;
narrow streak intersects hind-central eyes ; darkish line extends backwards from each lateral eye ; red thoracic indentation, enclosed by a broad, gamboge-coloured patch ; radial
striæ and marginal zone faintly defined by blackish streaks ;
two spots of a similar shade occur on each side on fore-half ;
almost glabrous ; oval ; pars cephalica large, nearly squarely
truncated ; lateral index less than facial ; convex, sides abrupt ;
pars thoracica convex, normal grooves slight ; medial indentation longitudinal ; profile-line rises from thoracic junction at
an angle of 60°, runs nearly horizontally to base of caput,
from thence slightly arched ; *clypeus* vertical, depth about
equal to space occupied by anterior central eyes.

Posterior row of *eyes* rather strongly procurved, of fair and
nearly equal size ; central pair on dark-brown oval spots, rather
the largest, and placed somewhat closer together ; anterior row
slightly recurved, of about equal size, less than two-thirds
smaller than hind-centrals ; median pair on dark rings, separated by an interval exceeding their diameter, scarcely one-
fourth more than that space from side-eyes ; laterals divided
by an interval surpassing breadth of a fore-eye, posited on
separate low, brown, tubercular eminences, posterior strongest.

Falces mahogany-brown ; transversely rugulose ; conical,
vertical, perceptibly exceed humeral joint of palpus in length ;
rather stouter than a femoral joint ; coarse black hairs.

Maxillæ linear, rounded, perceptibly truncated on superior side ; slightly inclined towards *labium*, which is oval,
apex emarginate ; rather narrow, one-half length of maxillæ ;
organs yellowish-brown, clouded with a deeper shade.

Sternum subovate, nearly as broad as long, sharply compressed, tapers between coxæ ; yellow-brown.

Legs, femoral, genual, and tibial joints yellowish, faint-green tinge ; about four not very clearly-defined annulations of a deeper hue on pars femoralis + tibialis ; metatarsi stone-colour, three brown annuli, darkest on hind-pairs ; legs slender, second and fourth of about equal length ; tarsal claws—first pair, strongly curved, 11 comb-teeth, form an even line from base to apex ; inferior claw sharply bent, 2 open teeth ; hairs sparse, spines rather strong.

Palpi, pars humeralis pale greenish-yellow, slightly exceeds cubital + radial joints in length ; cubital joint colour of humeral, viewed from above somewhat ovate, fore-end rather the widest ; radial joint light sienna-brown ; furnished with few long bristles ; more than one-third longer than former article, somewhat linear, bent into an obtuse angle ; projecting from base is a wide and long, viewed from above somewhat ear-shaped process, whose outer margin is armed with well-developed tooth-like serrations, basal strongest ; beneath its apex is a tumid, yellowish, somewhat semi-egg-shaped protuberance ; midway between the latter and base of process is a moderately stout, reddish apophysis, whose apex curves backwards ; lamina light-brown, ovate, pointed, strongly convex, moderately hairy ; genital bulb convex behind, deeply concave in front ; anterior surface chocolate-brown, extends nearly to terminal third of lamina ; posterior end yellowish-brown, pale within, somewhat membranous, prolonged forwards and inwards, on inner side, into a stout, moderately-curved process ; most remarkable appendages within bulbus are three fine black apophyses, upper curves backwards from extremity of bulb ; lower apophysis, basal half broad, membranous, curved forwards ; third apophysis of similar form, springs from inner side of bulbus ; directed forwards, above lower apophysis, is a broad, pale appendage drawn out on upper side into a black, strong, claw-like process, small tooth-like serrations ; suspended in front of this organ is a broad, membranous, pale-yellowish process, apex emarginate.

Abdomen oviform ; light yellow-brown, clouded with brown, dorsal band creamy colour, somewhat triangular-lanceolate.

Single specimen captured on Mount Egmont, *A. T. U.*

Gen. THERIDIUM, Walck.

Theridium punica-punctata, sp. nov.

Femina.—Ceph.-th., long, 1·4 ; wide, 1. Abd., long, 2·5 ; wide, 2. Legs, 1, 2, 4, 3 = 7·5, 5, 4·5, 3·2 mm.

Cephalothorax brownish-yellow, caput-grooves reddish ; hairs sparse ; pars cephalica convex, roundly truncated, lateral

index equals half facial; *clypeus* projecting, indentation below
eyes moderate, in height equal to two-thirds depth of ocular
area; pars thoracica oval, convex ; fovea oval, radial and caput
striæ fairly well defined ; profile-line somewhat level to posterior inclination, latter about 45°.

Eyes of tolerable and nearly equal size, on black spots;
posterior row moderately procurved, median pair rather closer
to each other— a space visibly exceeding an eye's breadth—than
they are to laterals; anterior row recurved, curvature perceptibly stronger than posterior line; centrals dark, smaller
than hind-pair, rather more distant from one another than
they are from hind-centrals, an interval slightly exceeding
their breadth; less than their diameter from side-eyes; hind-
lateral eyes about equal to anterior centrals in size, fore-eyes
smaller, posited obliquely on a common, lake-brown, tubercular eminence, contiguous.

Legs fulvous; hairs light, sparse; few slender black bristles;
moderately strong.

Palpi colour and armature of legs; slender, rather shorter
than cephalothorax.

Falces shade deeper than cephalothorax; sublinear, vertical; teeth form two rows on truncated apex, outer row 3
teeth, strongest projects from inferior angle of falx, inner composed of 5 small teeth.

Maxillæ strong, spathulate, inclined over *labium*, which is
wider than long, rounded, somewhat flattened at apex, less
than one-half length of maxillæ; organs light yellowish-
brown.

Sternum greenish-yellow, metallic reflections; lozenge-
shape, breadth equal to about three-fourths its length.

Abdomen oviform, convex ; ground - colour stone-brown
(possibly green tinge in fresh examples), dorsal band rather
broad, widens somewhat in centre, composed of numerous confluent creamy-white flecks; median streak vein-like, stained
with bright-yellow; in centre of dorsum are two large pinkish-
lake patches, separated by their own breadth; inferior half of
lateral. margins spotted with creamy-coloured flecks. *Vulva*
represents three equal-sized, closely-grouped foveæ, arranged
in triangle ; apical fovea above the rima genitalis represents a
brown rimed orifice ; hind-pair little more than brownish,
purple-margined circular figures.

Captured in the forest near Stratford, *A. T. U.*

Theridium apiatum, sp. nov.

Female.- Ceph.-th., long, 2; broad, 1·8. Abd., long, 3·4 ;
broad, 2·9. Legs, 1, 4, 2, 3 = 13·2, 9·3, 8, 7·3 mm.

Cephalothorax deep yellow-brown, median band marbled
with chestnut-brown, tinge of lake ; broad, as wide as eye-area

at hind-row, divided to fovea by a narrow streak; lateral band broad, lake-brown; few coarse hairs on caput; rather flatly convex, roundly truncated, lateral index equals space from a side-eye to hind-central furthest from it; *clypeus* inclined moderately forwards, scarcely equal to one-half facial space; pars thoracica oval, fairly convex; fovea subcircular, deep; normal grooves somewhat slight; contour of profile rounded behind, angle about 40°, slope forwards moderate, visibly curved.

Eyes tolerably large, on dark rings; enclose an oval space; posterior row of equal size, divided from one another by an interval perceptibly surpassing an eye's breadth; median pair their diameter and a half from fore-centrals; median eyes of anterior row one-fourth smaller than posterior pair, separated by their breadth and one-half; form with hind-centrals a quadrilateral figure widest in front; visibly more than their diameter from side-eyes; laterals contiguous, fore-eye rather the smallest; placed obliquely on a common, strong tubercle : have the pearl-grey lustre of posterior centrals.

Legs straw-colour; femora, patellæ, and tibiæ of first pair tinged with dark-orange; spotted—especially two first pairs —with lake; three brown-lake annuli on tibial + metatarsal joints of first and second; black-brown rings at extremity of three hind-pairs; first half of thigh of fourth pair black-brown; third and fourth have annuli on patellæ, and central and apical rings on tibial joints; similar but fainter annulations on metatarsi; hairs and bristles sparse.

Palpi pale-orange; hairs rather sparse; short bristles on cubital joint.

Falces fulvous, lake-coloured reflections at extremities; somewhat linear, vertical.

Maxillæ pale-yellow, suffused with orange; large, spathulate, tapering at extremities, inclined towards each other.

Labium yellowish, base greenish; nearly semicircular, somewhat truncated, less than half length of maxillæ.

Sternum pale greenish-yellow, metallic reflections, dark-lake flecks on side borders; triangular, about three-fourths as wide at base as long.

Abdomen oviform, convex; fore-third of folium without any determinate limits, gradually fades into ground-colour; centrally displays two pairs of crenatures, fore-pair twice as wide as hind; posterior third ovate; dull pale-brown (perhaps with an olive tinge in fresh examples), marbled with lake, border lake-black; median band broad, creamy-white on fore-third, centre part hastate, stained with lake and yellow; obliterated posteriorly; lateral margins shade lighter than folium, marbled with dull-purple; ventral region greenish-yellow, median band dull-lake : wide subtriangular mark of similar colour behind vulva. *Corpus vulvæ* black; oval, surrounded by two costæ rapidly widening in front, deeply cleft above the rima genitalis by a longitudinal, narrow-ovate, yellowish depression, on either side of depression is a broadovate fovea.

Stratford, *A. T. U.*

Theridium literatum, sp. nov.

Female.—Ceph.-th., long, 1·6; broad, 1·4. Abd., long. 3; broad, 2·3. Legs, 1, 2, 4, 3 = 8·2, 7, 6·4, 4·5 mm.

Cephalothorax amber-colour, median band on caput mahogany-brown, broad, somewhat crenate; marginal zone lightly pencilled with olive-brown, deeper tone on clypeus; few bristles; pars cephalica convex, roundly truncated, lateral index short; *clypeus* inclined rather forwards, height rather more than one-half facial space; pars thoracica broad-oval, convex; fovea elliptical, longitudinal; normal grooves fairly well defined; profile-line rises from thoracic junction at an angle of about 45°, slopes across caput with a slight curve.

Eyes large, on dark rings; posterior row nearly straight; centrals rather further from laterals than they are from each other, a space equal to three-fourths their diameter, removed from fore-centrals by about the former interval; anterior row moderately recurved; median eyes about one-fourth smaller than posterior pair, separated from one another by rather less than their diameter, and from side-eyes by an interval perceptibly shorter than their radius; laterals visibly larger than hind-centrals, have their pearl-grey lustre, contiguous, posited obliquely on moderately strong, dark, tubercular eminences.

Legs shade lighter than cephalothorax, lightly mottled, especially on thighs, with olive-green; two not strongly pronounced brown annuli on femoral, tibial, and metatarsal joints; basal rings on femora more or less obliterated; moderately slender; fairly furnished with hairs; few very slender spines; superior tarsal claws—first pair, 16 teeth, 14 rather even open comb-teeth, 2 basal small; free end moderately curved; inferior claw sharply bent, 2 points.

Palpi colour and armature of legs; humeral joint rather shorter than digital, and longer than genual and radial together; palpal claw 10 teeth, 7 strong. 3 basal smaller.

Falces amber-colour. fore-part mottled with olive-green, bare, semi-oval patch on base; conical, vertical, less than one-fourth shorter than digital joint of palpus, as stout as thighs of second pair of legs.

Maxillæ yellowish, large, oval, orange-brown patch in centre; stout, fore-half semi-elliptical, inclined towards lip.

Labium darker than maxillæ; semicircular.

Sternum yellow-brown, mottled with olive-green: clathrate; broad-cordate.

Abdomen oviform, depressed above, cone-shaped elevation on base; hairs moderately sparse, orange-red, spring from orange-red papillæ; integument greenish stone-colour; summit of cone whitish, fore and lateral margins clouded with olive-black; a large, somewhat W-shaped mark of same hue on posterior inclination above spinners, which have an orange colour; ventral surface normal colour; vulva yellowish, marbled with brown and olive-green; represents a large, somewhat quadrate projection; tapering forwards from basal angles is a moderately convex elevation—projecting rather beyond line of fore-angles, which are somewhat membranous—whose dark, obliquely-abscinded apex displays a heart-shaped fovea, apex of which is directed upwards.

Captured in the forest, Stratford, *A. T. U.*

ERYCINA, gen. nov.

Cephalothorax oval, lateral constriction at caput moderate, roundly truncated; profile-contour slopes forwards from limit of cephalic region, inclination more abrupt across ocular area, posterior incline moderately steep; *clypeus* slightly retreating, depth about equal to space between fore-centre eyes.

Eyes form two moderately-recurved rows, anterior strongest: posterior median pair and laterals of about equal size, posited on well-developed cup-shaped tubercular eminences; hind-centrals very perceptibly more distant from each other than they are from side-eyes; anterior centrals less than one-half size of laterals, seated obliquely on a low prominence, rather further from one another than they are from eyes next to them; laterals placed obliquely, interval between them perceptibly surpasses space dividing a fore-lateral from the hind-central nearest to it, an interval exceeding the distance between posterior median eyes.

Falces strong, second half divergent and somewhat attenuated, vertical, or slightly retreating.

Maxillæ rather longer than broad, taper to base, obtusely pointed.

Labium rather wider than long, margin very tumid, everted.

Sternum cordate: length scarcely exceeds the greater breadth.

Legs long, slender, 1, 2, 4, 3; hairs sparse; few fine bristle-like spines on the femoral, patellary, and tibial joints.

Palpi slender, digital joint longest.

Abdomen broad-oviform.

Erycina violacea, sp. nov. Plate XXI., figs. 4, 14, 17.

Mas.—Ceph.-th., long, 1·1. Abd., long, 1·7. Legs, 1, 2.
4, 3 = 6·9, 4·5, 4·2, 2·8 mm.

Cephalothorax pale yellowish-sienna; median band broad, olive-black, bifurcated on caput, limited by the hind-centre and lateral eyes; marginal zone rather wide, deeper shade; almost glabrous; clathrate; oval, lateral compression at caput slight; pars cephalica roundly truncated, convex, sides somewhat abrupt; *clypeus* perceptibly inclined towards falces, depth visibly exceeds space between fore-centre eyes; pars thoracica convex, indentation somewhat large and oval; caput and radial striæ faint; profile-line represents a slight curve across fore-end of caput, rises over a moderately prominent, rounded hump, from thence dips to thoracic junction at an angle of 60'.

Posterior median and lateral *eyes* of about equal size; hind-row moderately recurved, centrals seated on fair-sized tubercles, black oval spots; rather closer to side-eyes than they are to each other, an interval scarcely equal to an eye's breadth and a half, separated by about twice that distance from fore-centrals; anterior row more distinctly recurved; median pair rather smallest of eight, posited obliquely on a somewhat low elevation, their diameter and a half apart, less than that interval from side-eyes; laterals placed obliquely on cup-shaped tubercular prominences, space between them—which surpasses that dividing posterior centrals—rather exceeds the interval separating a fore-lateral eye from the hind-central nearest to it.

Falces light raw-sienna, slightly clouded; transversely rugulose; stouter than thighs of first pair of legs, somewhat linear, about twice as long as broad.

Maxillæ yellowish, green tinge, lightly clouded; length somewhat surpasses breadth, enlarged forwards, obtusely pointed, rather wide apart.

Labium fuscous; broader than long, margin very tumid, everted.

Sternum yellowish-brown, passing into brown about margins; broad-cordate, slight eminences opposite coxal joints.

Legs paler shade than cephalothorax, faint greenish annulations at apices of joints; hairs somewhat sparse; spines bristle-like, 3 or 4 on femoral + tibial joints; 1 on patellæ.

Palpi and legs concolorous; humeral joint somewhat incrassated forwards, rather exceeds in length cubital + radial joints together; pars cubitalis convex, perceptibly dilated in front; radial joint cup-shaped, slightly surpasses former article in length; pars digitalis large, lamina placed somewhat beneath bulb, fulvous, suffused with reddish-brown; hairs

fine, sparse ; broad-oval, basal half enlarged on inner side,
projects about midway to apex a flattish process of its own
colour, rather longer than wide, directed outwards and forwards ; two similarly-directed processes spring from basal
curve, inner linear, apex rounded, concave above, reddish
margins, extends to anterior process; outer acute, less than
half length of inner process; bulbus genitalis spiral (plane),
basal curve pale brownish-yellow, superior surface brownish-
amber, apical curve displays a conspicuous black beaded
margin, free end rather wide, tapering, blackish ; attached
to centre of bulb by a wide, veined, membranous keel extending nearly its entire length.

Abdomen oviform ; folium lyrate ; violet-drab, margins
dark-brown, broad fore-half displays a few somewhat silvery
flecks ; posterior half lightly suffused with a gamboge-
brown, flecks yellow, metallic ; constricted to about two-
thirds width of fore-part, very acutely-crenate ; superior streak
on lateral margins broad, green tinged gamboge-brown,
spotted with large silvery flecks ; inferior streak widest,
mottled with dark-brown, few metallic flecks ; shield on ventral surface light chocolate - brown, bordered with a fine
silvery line, three pale dots form a triangle near spinners.

Femina.—Ceph.-th., long, 1·5; broad, 1·2. Abd., long,
2·5 ; broad, 2·1. Legs, 1, 2, 4, 3 = 6, 4·8, 4, 2·5 mm.

Cephalothorax light raw - sienna colour, median band
broad, olive-black, extends from fovea to hind-pairs of eyes,
bifurcates on caput ; side-border greener hue ; few bristles ;
clathrate ; broad-oval, fairly compressed forwards ; pars cephalica convex, roundly truncated, sides rather abrupt, lateral
index equals three-fourths facial, shows two circular indentations placed transversely nearly midway between frontal line
and posterior limit ; pars thoracica convex, indentation at
posterior incline broad-oval, caput and radial striæ somewhat
slight ; profile-contour rises from thoracic junction at an angle
of 45°, perceptibly curved over occiput, slopes moderately
across ocular area ; *clypeus* directed slightly inwards, depth
scarcely equals space dividing fore-centre eyes.

Eyes on black oval spots ; posterior row only moderately
recurved, median eyes posited on strong cup-shaped tubercular eminences, rather more distant—twice an eye's diameter
—from each other than they are from laterals ; anterior row
more distinctly recurved, centre pair less than one-half size of
laterals, seated obliquely on a low prominence, about twice an
eye's breadth apart, somewhat less than that interval from
side-eyes, rather more than double their dividing-space from
posterior median pair ; laterals equal hind-centrals in size,
seated obliquely on well-developed cup-shaped tubercles,

rather more distant from each other than the fore-lateral eye is from the hind-central nearest to it.

Falces glossy, dark amber-colour, base clouded with olive-green; sub-conical, slightly retreating, second half somewhat divergent, nearly as stout as femoral joint of first leg, length slightly surpasses digital joint of palpus.

Maxillæ brownish-yellow, lightly clouded with olive-green; rather longer than wide, enlarged forwards, obtusely pointed, separated by an interval nearly equalling their own breadth.

Labium chocolate-brown; rather wider than long, margin tumid, everted.

Sternum yellow-brown, shading off to a deep-brown on margins; hairs sparse; cordate, length scarcely exceeds breadth between coxæ of fore- and hind-pairs of legs.

Legs colour of cephalothorax, femora faintly clouded and annulated with olive-green; two rings on tibiæ + metatarsi of a brownish hue; hairs sparse, outstanding; spines bristle-like; 3 or 4 on femora; patellary joints, 1; tibiæ of anterior pairs, 4; metatarsi, 1; tibiæ of hind-pairs, 2 spines; metatarsi, 1.

Palpi yellowish, clouded with olive-green; slender, rather shorter than cephalothorax; armature bristle-like hairs, few slender spines; palpal claw long, slender, slightly curved, free end more than half length of claw, 6 comb-teeth increasing in length.

Abdomen broad-oviform, projects over base of cephalo-thorax, viewed laterally somewhat reniform; ground-colour yellowish-drab, faintly suffused with purple-brown approximating to violet, flecked with more or less oval creamy-coloured spots, bordered with violet-brown; folium sublyrate, margins interrupted, brown, yellow outer border, fore-half nearly twice as broad as posterior; ventral field and inferior half of lateral margins greenish-black, figured with oblique marks and free and coalescing spots; shield lozenge-shape, broad, sides sinuating; yellowish, passing into pale-purple on margins, centre clouded with dark-olive.

Vulva yellow-brown, somewhat ⊥-shaped, moderately elevated, transversely rugulose; displays two small, reddish, elliptical foveæ, their greater diameter intersected by a septum whose breadth nearly equals their transverse diameter, septum bordered by dark costæ, which rapidly diverge round foveæ, extend outwardly above the rima genitalis, following somewhat the lower margin of two large, convex, oval, transverse depressions, contiguous to foveæ.

This species was not uncommon in the forest near ·Stratford, male examples by far the most numerous. *A. T. U.*

Gen. CORNICULARIA, Menge.

Cornicularia crinifrons, sp. nov. Plate XXI., fig. 3, 11.

Mas.—Ceph.-th., long, 1·4; broad, 1. Abd., long, 2·8; broad, 1. Legs, 1st pair missing, 2, 4, 3 = 10·6, 9, 5·5 mm.

Cephalothorax fulvous, eye-area red-mahogany colour, shaded with brown, sharply defined, connected by a wide streak with a brown-bordered lanceolate figure extending to fovea, latter reddish; marginal zone narrow, dark-brown; middle band represented by a series of dots on radial ridges; cephalic tubercle furnished with numerous strong hairs, directed outwards and forwards; pars cephalica elevated, laterally rounded, prolonged into a strong tubercle, apex rounded, rather more than one-third length of caput; *clypeus* directed moderately forwards, height nearly equals depth of ocular area; pars thoracica nearly circular, convex, border-hem wide, fovea circular, deep; radial striæ deeply grooved, caput striæ moderately: contour of profile slopes backwards from summit of caput to posterior limit at an angle of about 30°, with a visible curve, dome-shaped on thorax.

Eyes do not differ much in size, form a circlet on cephalic eminence; posterior eyes of about equal size, centrals closer to each other—a space less than twice an eye's breadth—than they are to laterals; anterior row equidistant, median pair dark, smallest of eight; laterals have the pearl-grey lustre of hind-median eyes, posited obliquely, contiguous.

Falces brownish-amber, glossy; subconical, anterior fourth divergent, directed somewhat forwards, moderately stout, length rather less than twice height of clypeus.

Maxillæ fulvous, pale about extremities; rather longer than broad, obtusely pointed, strongly developed at insertion of palpi, inclined over *labium*, latter organ fulvous, breadth somewhat surpasses length, apex pinched, lip-like, less than half length of maxillæ.

Sternum fulvous, broad, well-defined heart-shape.

Legs light brownish-yellow; femora of second, third, and fourth pairs have respectively 5, 3, and 4 light-brownish rings, tibiæ 4, 2, 2; legs very slender; hairs sparse, few fine bristles; superior tarsal claws—second pair, long, rather slender, moderately curved, 5 or 6 small teeth, 1 long, strong tooth, free end forms nearly half the claw, tip bent; inferior claw three-fourths length of superior, sharply bent, base stout; apparently small teeth.

Palpi colour of legs; humeral joint surpasses cubital + radial by one-fourth; contour of cubital joint has somewhat the form of an isosceles triangle; on apex—which projects above its articulation with penultimate article—is a stout, curved, black spine; two serrations on central third, anterior side;

moderately long bristle behind; radial joint linear, fully
equals the greater length of former article; laminæ bulbi
obovate, directed towards each other; hairs sparse; margin
at base, outer side, turned up, forming an oval, deep depression above; bulbus genitalis dark amber-colour, well-developed,
convoluted; convolutions terminate in two large well-defined
apophyses directed downwards, of about equal length; inner
apophysis lake-colour, dark margins, broad, rounded, outer
face concave, projects laterally two small processes; back-
ward-directed process black, horn-like; fore-process yellowish,
curved forwards and upwards, apex tumid; outer apophysis
spiral (single curve), base tumid, free end stout, acute,
blackish.

Abdomen elongate - oviform; creamy - brown, somewhat
mottled with brown-black; mottling composed of small elongate
patches interspersed amongst free and confluent spots; dorsal
region somewhat free from marks; ventral surface light-
brown; spinners yellowish.

Single example, Stratford, *A. T. U.*

Fam. EPEIRIDÆ.

Gen. EPEIRA, Walck.

Epeira atri-apiata, sp. nov.

Mas.—Ceph.-th., long, 4·3; wide, 3·5. Abd., long, 5; wide,
3·5. Legs, 1, 2, 4, 3 = 15·5, 14, 10, 8 mm.

Cephalothorax rich mahogany-brown; glabrous, except few
white hairs about clypeus; closely studded with small papillæ;
sharply compressed forwards; pars cephalica depressed, central indentation, ocular elevation projects prominently over
falces; depth of *clypeus* rather more than half facial space;
pars thoracica strongly convex, median indentation oval, large,
deep, discloses a cruciate groove within; radial and caput striæ
more or less obliterated; contour of profile represents a low
arch rising at eye-prominence.

Fore- and *hind-row* of *eyes* recurved; posterior median
pair, on dark patches, about one-third smaller than anterior
centrals, divided by an interval equalling twice their own diameter; fore-median pair separated from each other by an
eye's breadth and a quarter, and from hind-centrals by fully
their diameter; laterals one-third smaller than posterior median pair, posited obliquely on a dark common tubercle.

Falces fulvous, suffused with olive-brown, moderately stout
and long, second half curved upwards and outwards. ·

Maxillæ rather longer than broad, roundly pointed, inclined towards *labium*, latter perceptibly wider than long,

rounded, scarcely half length of maxillæ; organs olive-black, yellowish tips.

Sternum glossy black-brown; cordate.

Legs yellowish-brown; three-fourths of femoral joints of two first pairs, and basal half of hind-pairs, mahogany-colour; patellæ, central and apical rings on tibiæ + metatarsi, and second half of tarsi, similar shade; almost glabrous; spines brown, rather short, fairly numerous, irregular; coxæ of first pair armed with a curved process; tibial joints of first legs cylindrical, spines somewhat in rings; tibiæ of second stouter, enlarged on inner side at base and third quarter, spines strong, especially distal groups.

Palpi fulvous, olive-green tinge; hairs whitish; pars humeralis short, incrassated forwards; cubital joint one-third shorter than former article, broad-oval, projects a bristle from extremity; radial about as long as cubital, broad, cup-shaped, upper side emarginate; pars digitalis large, subglobose; lamina fulvous, clouded with olive-green; hairs fine, short, sparse; elongate-ovate, depressed, apices curved backwards; base tumid, centrally intersected by a dark groove; outwardly prolonged into a triangular area terminating in a moderately short, down-curved, lake-coloured apophysis; genital bulb viewed somewhat from front displays, including cap, five lobes; bulbus cap reddish-brown, golden reflections, partially encircled and indented by a conspicuous black apophysis; first lobe beneath cap narrow, encloses face and inner side of bulbus, yellow-orange metallic reflections, two transverse brown streaks; second lobe largest, side deeply emarginate, projecting from centre of this cavity is a short, slightly-curved, black process, directed backwards; transversely rugose, bright crimson-lake, passing into yellowish and black about the vertically-rugulose border; lower lobe light-brown, yellowish metallic reflections, rugose, wide margins transversely wrinkled, viewed beneath outer extremity reclinate; projecting from beneath former and contiguous to latter lobe is a broad sharply-backward-curved process, curvature at extremity of lobe, apex squarely truncated, light-brown, yellowish reflections, black-brown on outer side, lake band at curvature; beneath a yellow-brown, large, reniform lobe at base of bulb is a wide, backward-curved, yellowish-brown process, base tumid, apex rounded, revolute.

Abdomen oviform, subdepressed, humeral tubercles slight; three first and fifth tubercles very short; tubercle of second row prominent; stone-colour, lightly suffused with slate; folium stained with pale-brown, sparingly dappled with rather large slate-coloured lobate spots; margins acute-crenate, brownish, fore- and hind-extremities obliterated; four brown, conspicuous, impressed spots form a trapezoid narrowest in front; lateral margins flecked and figured with slate-colour: ventral region occupied by an olive-green shield, yellow transverse streak above.

Single specimen, contained in *Mr. H. Suter's* collection from Hastwell, Forty-mile Bush.

Epeira acincta, sp. nov.

Femina.- -Ceph.-th., long, 4 ; broad, 3·3 ; facial index, 1·7. Abd., long, 6·4 : broad, 5·3.   Legs, 1, 2, 4, 3   15, 14, 12. 9 mm.

Cephalothorax brownish-orange, very lightly suffused with olive-green ; hairs sparse, pale orange-yellow ; slightly shorter than patella + tibia of first leg ; cephalic part moderately convex, sides abrupt, eye-prominence rather low ; lateral index equals space between fore-lateral eyes ; *clypeus* inclined forwards, depth scarcely equals diameter of a fore-central eye ; pars thoracica moderately convex, sides rounded ; fovea somewhat oval, normal grooves rather faint ; contour of profile rises with slight curve at an angle of nearly 40° from peduncle, moderately inclined forwards, dips rather abruptly across ocular prominence.

Eyes on dark rings ; posterior row slightly recurved, hindmargin of centrals in line with fore-margin of laterals of same row ; median pair visibly larger than fore-centrals, interval between them slightly exceeds an eye's breadth, less than their space and one-half from laterals ; anterior row more distinctly recurved, centrals rather more distant from each other—an eye's diameter and a half—than they are from posterior centrals, perceptibly more than their space from side-eyes : laterals about one-third smaller than anterior centrals, posited obliquely on low tubercles, rather more than their radius apart.

Falces shade lighter than cephalothorax ; conical, vertical, somewhat gibbous at base in front ; one-fourth shorter than radial + digital joints of palpus, scarcely equal to femur of second leg in stoutness.

. *Maxillæ* brown-yellow, base suffused with light-brown ; nearly as wide as long, dilated forwards, obliquely truncated, inclined over *labium*, base of latter organ chocolate-brown, apex pale greenish-yellow ; oval, length nearly equals breadth.

Sternum dark chocolate-brown, yellow, sublanceolate central mark ; cordate, eminences opposite coxal joints.

Legs and cephalothorax concolorous ; thighs of first pair lightly suffused with lake on second half ; patellæ + tibiæ tinge of green ; faint indications of greenish annuli on metatarsi + tarsi ; femora of third legs visibly marked with two olive-green rings ; patellar joints of hind-pairs greenish ; faint annulations of same hue on other articles. Legs moderately strong, tibiæ as long as metatarsi; hairs sparse, yellowish; spines yellow, base dark; first pairs, about 15 on femora; few on patellæ; about 20 spines—equalling diameter of article in length—on tibiæ; 7 on metatarsi; 3rd + 4th femoral joints, 5; patellæ, 3; tibiæ of 3rd 7, of 4th 9 spines; metatarsal joints of third pair 3, of fourth pair 14; tarsal claws of first pair 8 teeth, 4 small, rest stout; inferior claw, 2 close teeth.

Palpi colour and armature of legs, green rings at apices of cubital + radial joints; moderately slender, equal cephalothorax in length; palpal claw, 8 teeth.

Abdomen triangular-oviform, depressed, humeral tubercles very slight; ground-colour dull pale-green, thickly dappled with stone-colour, resolved into oval spots on base; folium subtriangular, margins subcrenate; lightly suffused with pale-drab; margins evanescent, marked out by a series of five or six brown dots; basal end defined by procurved angular row of large dots; four dark impressed spots form a trapezoid, narrowest in front. *Vulva* greenish amber-colour, shading off to dark-brown on side-margins; transversely wrinkled; tolerably convex, compressed rapidly into a large tapering scapus, whose length is about equal to the greater diameter of organ: stylus displays about two wrinkles, apex represents a broad conical cap, deep obtuse notch in upper margin, twice as wide as long, measured from apex of notch; scapus bordered by broad tumid costæ, showing three or more distinct grooves.

Captured at the base of Mount Egmont, on *Rubus australis*. This climbing shrub, which is generally to be met with about the skirts of forests, is much frequented by spiders, its close foliage and prickly armature offering the double advantage of protection to themselves and their prey from their mutual enemies, birds. This species belongs to a rather handsome group of *Epeira*, which is to be met with throughout New Zealand.

Epeira nigro-hastula, sp. nov. Plate XXI., fig. 13.

Mas.—Ceph.-th., long, 2·4; broad, 2. Abd., long, 3: broad, 2·2. Legs, 1, 2, 4, 3 = 10·5, 8·5, 7·8, 5 mm.

Cephalothorax fulvous, passing into green on caput, whose yellowish reflections disclose a somewhat intricate pattern; brown-black hastate figure within fovea, lance-head reaches limit of caput, somewhat removed from but forming a transverse line with its apex are two brown spots; hairs whitish, sparse; ovate, moderately compressed forwards; cephalic region flatly convex, ocular prominence fairly developed, lateral index fully equals facial; thoracic fovea oval, longitudinal: normal grooves moderately deep; profile-line represents a rounded arch, perceptible obtuse eminence on occiput.

Posterior row of *eyes* tolerably recurved, centrals equally distant from each other and fore-centrals, an interval equalling an eye's breadth and a half; rather less than their space from laterals; anterior row recurved; median pair one-third larger than hind-pair, an eye's diameter apart, scarcely that interval from side-eyes; laterals about one-third smaller than posterior median eyes, seated obliquely on brown, moderately prominent tubercular eminences, separated by half their radius.

Falces remarkably slender, yellowish, green shade; transversely rugulose; length fully equal to breadth of posterior row of eyes, scarcely as stout as tibia of second leg; inclined inwards, taper moderately, fore-third bent outwards and upwards; base shows a tumid collar; apices rest within concavity of maxillæ.

Maxillæ rather longer than wide, roundly pointed, directed inwards, prolonged beyond falces.

Labium longer than broad, triangular; organs pale greenish-yellow, base brownish.

Sternum green; cordate, slight eminences opposite coxal joints.

Legs pea-green, fuscous shade on superior surface of thighs, interrupted dark annuli beneath; moderately slender; coxæ of first legs project a curved process; tibiæ linear; hairs sparse; spines yellowish, brown apices; femora of first pair 5 spines on superior surface, cluster of 6 long spines beneath; patellæ, 5; tibiæ, 12 scattered spines; metatarsi, 3; spines of second pair shorter than first, of nearly equal number; spine armature of two hind-pairs does not differ greatly, more sparse than fore-pairs.

Palpi, humeral + cubital joints greenish, former short; cubital broad-ovate, projects two long reddish bristles; pars radialis duller green, shorter than cubital joint, prolonged on outer side into a large, semi-pellucid, subconical process, tumid on inner side, furnished with few bristles; laminæ bulbi yellowish, clouded with dark-green, moderately hairy; ovate, tapering, directed towards each other; base of lamina prolonged into a rather wide and long reddish process, directed outwards, apex curved; bulbus, viewed from somewhat behind and beneath, discloses several remarkable appendages; basal olive-brown, stout, subcylindrical, somewhat crescent-shaped, curved downwards; inner half tapering, apex cleft into two processes, outer broadly pointed, displays a series of short teeth; inner process longest, acute; outer extremity of lunulate process broad, apex fin-like, fore-margin furnished with a fringe of black, very acute projections, increasing in size upwards round apex; hind-angle prolonged, somewhat claw-like, projects a short tooth near its base; projecting forwards from margin of bulb is a wide, oblong, yellow-brown, hood-like membrane, with brown transverse stripes and reddish beaded margins; immediately beneath the latter is a paler projection of nearly equal length, its tumid fore-margin armed with three convergent rows of stout black teeth; directed downwards and forwards from centre of bulbus is a large olive-brown process, base constricted; base of the subtriangular fore-part somewhat spiral, apex membranous; curving inwards from front of genital bulb is an easily-perceptible, semi-pellucid, cylindrical apophysis, dilated at apex. Viewed from front, suspended beneath apex of lamina is a reddish, broad, membranous process, concave on outer side, fore-half constricted, inner angle drawn out, acute.

Abdomen oviform, moderately convex; somewhat sparse bristle-like hairs projecting from brown dots; folium sinuate, base semicircular, projects a short petiole; grey-green, margins green; enclosed within is a figure of similar form, pea-green, border deeper shade; the sublanceolate pale median mark projects backwards from the constriction at the almost circular basal end; lateral margins green, obliquely streaked with dark-green; ventral field brownish, shield shows two dark spots, border light flecks.

Femina.—Ceph.-th., long, 2·6; broad, 2·1. Abd., long, 4·9; broad, 4·3. Legs, 1, 2, 4, 3 = 10, 8·2, 8, 6 mm.

Cephalothorax pea-green, fading about thoracic region; yellowish pattern on caput, more conspicuous in spirit; black hastate figure in fovea; hairs pale straw-colour, mostly on cephalic part; pars cephalica tolerably convex, sides abrupt, lateral index equal to space occupied by a hind-lateral eye and the central furthest from it of same row; eye-eminence somewhat prominent; height of *clypeus* equal to diameter of a fore-median eye; pars thoracica ovate, convex; indentation rather deep, longitudinal; striæ tolerably well marked; contour of profile represents an angle of 50° on posterior slope, inclined moderately across occiput, abrupt dip at ocular area.

Posterior row of *eyes* moderately recurved, centrals posited somewhat obliquely, visibly more than an eye's breadth from each other and fore-centrals, nearly their space from laterals; anterior curvature rather stronger than posterior; median eyes about one-fourth larger than hind-pair, separated from one another by a distance perceptibly more than an eye's diameter, somewhat more than that interval from side-eyes; laterals about one-fourth smaller than hind-median, placed obliquely on dark tubercles, half their radius apart.

Falces yellowish-green; conical, moderately tumid at base in front, vertical, rather slighter than a thigh of second leg, length equals breadth of facial space.

11

Maxillæ rather longer than broad, pointed, base slender, tolerably inclined towards each other.

Labium scarcely as long as wide, roundly pointed, everted ; organs brownish, apices greenish-yellow.

Sternum yellowish, green tinge ; broad-cordate, eminences opposite coxal joints.

Legs yellowish-green, faint annulations ; hairs sparse ; spines yellowish, dark points, short ; femora of first leg, 3 spines above, 5 beneath ; patellary joint, 4 ; 10 on tibia, about equal to diameter of article in length ; metatarsus, 3 ; spine armature of second leg about equal to first ; femoral joints of hind-pairs have no spines on inferior surface, otherwise their spine armature nearly equals anterior pairs.

Palpi coloration of legs ; hairs only ; slender, nearly equal to cephalothorax in length.

Abdomen oviform, somewhat depressed, moderately clothed with bristle-like hairs ; folium sinuate, base rounded, petiole short ; closely flecked, pale grey-green, border green ; encloses a similar figure of a deeper shade ; median band wide, lighter tone ; red-brown spots at root of hairs. *Vulva* represents a yellowish, transverse, broad-oval, depressed area, wrinkled in circles ; free part or scapus sharply bent, more than half of area, projects somewhat over the rima genitalis, superior margin scalloped, lateral segments narrowest ; displays in centre a rather large, lake-brown, rugose spot ; side-margins of corpus involute ; bordering the rima genitalis is a dark-lake narrow costa, extremities tumid, revolute, slightly surpasses centre segment in length.

Taken in the forest near Stratford, *A. T. U.*

Epeira atri-hastula, sp. nov. Plate XXI., fig. 7.

Femina.—Ceph.-th., long, 2·8 ; broad, 2·1 ; facial index, 1. Abd., long, 4·8 ; broad, 4·6. Legs, 1, 2, 4, 3 = 11·5, 10, 8·5, 6·5 mm.

Cephalothorax, caput green, fading off to a paler hue on thorax ; cephalic part—especially after turning yellow in spirit—displays an intricate design ; conspicuous brown-black hastate figure on thorax, transverse base-lines across fovea, lance-head intersects caput striæ, whose posterior halves are defined by brown streaks ; somewhat removed from and in line with apex of lance-head are two dark spots ; moderately furnished with light hairs on cephalic part ; pars cephalica rather convex above, sides abrupt, roundly truncated, ocular eminence fairly prominent ; lateral index nearly equals facial ; depth of *clypeus* equals diameter of a fore-central eye ; pars thoracica broad-oval, convex ; fovea oval, longitudinal ; normal grooves moderately strong ; profile-contour rises at an angle of 40°, moderately inclined forwards, dips abruptly from hind-row of eyes.

Eyes on dark rings, subequal; posterior row visibly re-curved, interval between median pair equal to rather more than an eye's breadth, visibly more than their space from laterals; anterior row more distinctly recurved, centrals perceptibly exceed hind-pair in size; rather further from each other than they are from posterior centrals, a space slightly surpassing an eye's diameter; laterals rather the smallest of eight, posited obliquely, half their radius apart, on low tubercles.

Legs yellowish-green, faint annuli; hairs pale-yellow, somewhat sparse; spines yellow, dark base; femoral joints have few short spines on superior surface; first pair has also a row of 4 on inferior side; second pair, 2; patellæ, 3 or 4; tibiæ of first, 9; three hind-pairs more sparsely spined; meta-tarsi, 3 or 4.

Palpi pale-green; cubital joint projects a bristle; fully equal to cephalothorax in length.

Falces yellowish-green; conical, moderately gibbous at base in front, fore-end somewhat divergent; length equals tarsus of a fore-leg, scarcely as stout as femur of second leg.

Maxillæ brown, apices pale-green; length somewhat sur-passes breadth, roundly pointed, taper to base; moderately inclined towards each other.

Labium colour of maxillæ; rather wider than long, pointed.

Sternum greenish; broad-cordate, eminences opposite coxæ.

Abdomen broad-oviform, moderately convex above; hairs sparse; pale grey-green, closely flecked; folium undulating; deeper green, few red-brown spots; ventral shield lozenge-shaped; olive-brown, broad margin of creamy-coloured flecks. *Corpus vulvæ* olive-brown; strongly-wrinkled, transverse, oval eminence; tumid and longitudinally depressed above the rima genitalis; scapus green-tinged, bright reddish-amber; projects from tumid eminence, long, base broad, vermiform, flattened, curves forwards on abdomen for half its length, sharply bent backwards to little beyond posterior limit of corpus; apex semi-oval, somewhat calceolate.

Forest, Stratford, *A. T. U.*

Epeira galbana, sp. nov.

Femina.—Ceph.-th., long, 2·2; wide, 2. Abd., long, 4·1; wide, 3·4. Legs, 1, 2, 4, 3 = 9·4, 8·5, 7·1, 5·5 mm.

Cephalothorax brownish amber-colour, V-shaped yellowish mark at limit of caput; hairs whitish, soft, sparse; oval, moderately compressed forwards; pars cephalica convex, roundly truncated, ocular prominence fairly developed, lateral index scarcely equals facial ; height of *clypeus* slightly exceeds diameter of a fore-centre eye ; pars thoracica convex, fovea circular, deep, large ; normal grooves somewhat slight ; contour of profile represents an angle of 45° at posterior slope, visibly curved across caput.

Eyes moderately recurved, curvature of posterior row nearly as strong as anterior; median eyes of hind-row large, on dark oval spots, separated by an interval equal to their diameter and a quarter ; three-fourths their space from laterals ; fore-centrals on dark spots, nearly form a square with posterior median pair, scarcely equal to one-half their size, divided from them by a space less than their own breadth ; closer to side-eyes than are hind-centrals; lateral eyes one-fourth smaller, than fore-median pair, posited obliquely on separate dark tubercular prominences, less than one-third their diameter apart.

Legs pale raw-sienna; femora + patellæ suffused with a deeper shade ; two faint annuli on tibiæ + metatarsi ; hairs sparse ; rather slight black spines on all joints except tarsal ; femora of first have 11 ; patellæ, 4 ; tibiæ, 14 somewhat irregular spines ; metatarsi, 6 ; superior tarsal claws—first pair, 10 somewhat even comb-teeth ; inferior claw rather sharply bent, 2 points ; legs moderately slender.

Palpi, paler hue than legs ; armature similar ; equal cephalothorax in length ; palpal claw straight, free end moderately curved ; 8 rather even comb-teeth.

Falces and cephalothorax concolorous ; conical, convex, vertical, perceptibly surpass cubital + radial joints of palpus in length, somewhat slighter than femur of a fore-leg.

Maxillæ yellow-brown, clouded ; nearly as broad as long, taper moderately to base ; inclined towards each other.

Labium yellowish-brown, base chocolate-brown ; rather wider than long, pointed.

Sternum light-brown, centre stripe broad, undulating. yellow ; cordate, eminences opposite coxal joints.

Abdomen oviform, base pointed, projects over cephalothorax; hairs fine, sparse ; dorsal field pea-green, few scarlet dots ; anterior pair of the four well-marked impressed spots olive-green, connected by a somewhat angular mark of a similar hue ; almost in line with posterior pair is a semicircular olive-coloured figure, which throws off a series of nearly parallel yellowish vein-like lines, extending to spinners ; lateral margins yellow, numerous brownish nearly vertical lines curve up-wards from ventral surface ; shield light chocolate-brown, margin defined by yellow flecks ; on face of shield are two broken parallel yellow bars. *Vulva* yellow-brown, green tinge ; corpus rapidly developed into a large, flatly convex triangular scapus, transversely wrinkled, rather wider than long; its rather broad beaded margins, which project upwards and outwards, are sharply bent inwards at their extremities; the confluent beading forms a semicircular wrinkle at base of broad stylus, which rather exceeds the dark-stained angular projections in length; base of stylus consists of two wrinkles; mouth of large ascidium-shaped apex inversely ovate; few coarse hairs.

Forest near Stratford, *A. T. U.*

Epeira venustula, sp. nov.   Plate XXI., fig. 12.

Femina.—Ceph.-th., long, 2; wide, 1·8. Abd., long, 3·5; wide, 3.   Legs, 1, 2, 4, 3 = 8, 6·3, 5·1, 3·6 mm.

Cephalothorax burnt-sienna colour, speckled in radiating lines with a deeper hue; triangular area, including facial space, of a gamboge-yellow, pale metallic reflections, extends to limit of caput; clathrate; hairs light-yellow, very sparse; length equal to patella + tibia of a leg of second pair; pars cephalica convex, lateral index nearly equals facial; eye-prominence moderate; *clypeus* retreating, height scarcely equal to radius of a fore-central eye; pars thoracica moderately convex, sides well rounded; fovea suboval, deep; caput and radial striæ fairly defined; contour of profile rises from thoracic junction at an angle of 40°, arched across cephalic part, dips rather abruptly at ocular area.

Eyes on dark spots, recurved, first rather the strongest curve; posterior row divided by an interval rather exceeding their diameter; scarcely more than their space from laterals; anterior centrals about an eye's breadth and a half from each other; their diameter from hind-pair, which they equal in size; separated from side-eyes by an interval equal to three-fourths their space; laterals about half size of centrals, posited obliquely, one-third their diameter from each other, on separate moderately strong tubercles.

Legs brownish creamy-colour; femoral joints have wide brown—dark-brown on hind-pairs—annuli on fore-half, passing into greenish-yellow at extremities; patellæ suffused with reddish-brown; tibiæ of first + second show two free, faint, reddish-brown rings; tibiæ of hind-pairs, and metatarsal joints of all, have central and apical annulations; legs somewhat slight; hairs sparse; few black spines.

Palpi yellowish, ring on extremity of radial and fore-half of digital joints red-brown; hairs sparse.

Falces fuscous, clouded with olive-green; conical, gibbous in front, vertical, stouter than thigh of a fourth leg, length equal to breadth of anterior row of eyes.

Maxillæ light greenish-brown, base chocolate-brown; about as wide as long, roundly pointed, taper to base, moderately inclined towards each other.

Labium greenish-black, apex greenish-yellow; somewhat triangular, rather wider than long.

Sternum chocolate-brown, median mark lanceolate, yellowish; cordate well-defined eminences opposite coxæ.

Abdomen broad-oviform, flatly convex above, projects well over base of cephalothorax; dorsal field covered by a large, acutely-sinuate, leaf-shaped figure; petiole formed by two Ɔᴐ-shaped marks; ground-colour creamy-white, except basal end almost entirely suffused with yellowish-olive, spotted with numerous creamy-coloured, lake-margined flecks; fore-part shows few lake and yellow spots; border stained with greenish-yellow; three pairs of blackish impressed spots, centre pair furthest apart, placed in deep indentations; folium tapers moderately to spinners, somewhat acute-crenate, truncated base intersects and extends beyond centre impressed spots, defined by a rather broad, procurved, soft brown-black band, intersected by posterior end of the somewhat diamond-shaped figure extending along its anterior margin; latter mark creamy-white, lake spots, yellowish-olive centre; border of folium pale greenish-yellow, crenatures brown-black; area is of the normal olive-green, and, like dorsal surface, shaded with a deeper hue, prettily marked with pale-yellow, creamy, and lake coloured sinuate lines and spots; lateral margins yellowish-olive, passing into a lightish-brown on ventral surface, marbled with deep greenish-black; shield somewhat oval, includes spinners; cinereous, flecked; above two large centre yellow spots are two lateral, blackish, ovate marks; border creamy, broken into four dots in line with spinners. *Vulva* yellowish amber-colour; represents an elliptical, centrally-compressed, smooth lobe; wrinkled, tumid margins somewhat Ɛꓢ-shaped, dark spot in short upper curve; scapus large, somewhat spoon-shaped, apex rather deep, curves over lobe, projects rather beyond.

Single example, taken in the forest near Stratford, *A. T. U.*

Epeira melania, sp. nov.

Femina.—Ceph.-th., long, 4·3; wide, 3·6; facial index, 1·9. Abd., long, 5·7; wide, 4·5. Legs, 1, 2, 4, 3 = 18, 16, 14·5, 9·5 mm.

Cephalothorax fulvous; hairs pale, sparse; length equals patella + tibia of a fourth leg; pars cephalica moderately convex, truncation perceptibly curved, ocular elevation slight; lateral index about equal to facial; depth of *clypeus* equals diameter of a fore-centre eye; slight lunulate procurved indentation behind posterior centre eyes; pars thoracica moderately convex, sides fairly rounded; transverse indentation on posterior inclination; caput and radial striæ tolerably strong;

profile-line rises from thoracic junction at an angle of 30°, slopes somewhat steeply with a slight curve to edge of occiput.

Eyes small; posterior row perceptibly recurved, median pair separated by an interval visibly surpassing their diameter, divided from laterals by their space and a half; anterior row recurved, centrals dark, black rings; largest of 8; removed from each other and hind-centrals by a distance scarcely equal to twice an eye's breadth; rather more than their space from side-eyes; laterals have the lake tinge and lake-brown rings of hind-centre eyes, placed obliquely on a low tubercle, their radius apart.

Legs fulvous, green tinge on thighs; dark patch beneath patellary joints of two first pairs; somewhat obscure brown annulations on metatarsi; hairs light, sparse; few light-yellowish spines on femora, cluster of strong spines on inner side of first pair; spines moderately strong and numerous on other joints; superior tarsal claws—first pair, 9 open teeth, increasing in length and strength; inferior claw, 2 close teeth.

Palpi pale brownish-yellow; armature of legs; moderately slender, length of cephalothorax; palpal claw moderately curved, 6 open teeth.

Falces shade paler than palpi, glossy, conical, gibbous in front, stouter than femur of first leg, length equal to digital joint of palpus.

Maxillæ light-brown, green tinge, margins paler; nearly as broad as long, roundly pointed, inclined over *labium*, latter organ shade darker; wider than long, subtriangular.

Sternum fulvous, marbled with green; cordate, eminences opposite coxæ.

Abdomen triangular-ovate, depressed above; folium subtriangular, acutely-crenate; two large, black, nearly contiguous spots on humeral angles; small black spot on either side of petiole, which is broad, curves round base of abdomen, black, white spot; ground-pattern consists of a series of creamy-coloured, purple-spotted, sinuous lines, converging upwards, resolved more or less into coalescing spots, intersected by purple streaks; lightly suffused with olive-brown; folium suffused with a deeper shade; crenatures on posterior end brown-black; ventral field purple, sides brown; shield quadrate, light olive-brown, border flecked. *Vulva* light greenish-brown, passing into olive-green on lateral margins, stylus yellowish amber-colour, side-lobes glossy brown-black; *corpus vulvæ* projects outwards, rather wider than long, strongly convex, transversely wrinkled; scapus large, about length of corpus, transversely rugose, sides developed into two very remarkable centrally-depressed discs, surpassing shaft of scape in length; deep sides of discs project outwards at an obtuse angle from

corpus; stylus short, broad, rugose, apex large, well-formed
semi-globose cup, with a conspicuous amber-brown beaded
rim, latter in line with fore-margin of discs.
Stratford, *A. T. U.*

Epeira similaris, sp. nov.

Femina.—Ceph.-th., long, 4 ; broad, 3·1 ; facial index, 1·5.
Abd., long, 7·5 ; broad, 6·5.   Legs, 1, 2, 4, 3 = 14·4, 14, 11,
8 mm.

Cephalothorax dull raw - sienna, cephalic parts lightly
clouded with olive-brown ; hairs sparse, white and silky on
thorax, mostly of an orange-yellow on caput ; length equal to
tibia of first pair; pars cephalica moderately convex ; ocular
eminence fairly prominent ; lateral index nearly equals facial ;
sides of pars thoracica well-rounded ; indentation moderately
deep, caput and radial striæ not strongly defined ; profile-line
represents a not very strongly-curved arch ; *clypeus* directed
inwards, depth equals breadth of a fore-central eye.

Fore- and hind-row of *eyes* nearly equally recurved ; anterior centrals perceptibly smaller than posterior pair, separated
from each other by an interval equal to rather more than an
eye's breadth, and from hind-pair by a space scarcely equalling
the former ; more than two-thirds their space from laterals ;
median pair of posterior row visibly more than their diameter
apart ; their space and a quarter from side-eyes of same row ;
laterals posited obliquely on separate, low, dark, tubercular
eminences, perceptibly less than their radius from each other.

Legs, thighs of two first pairs brownish-yellow, faint indications of annuli ; patellæ deeper shade ; tibial, metatarsal,
and tarsal joints greenish tinge passing into light - brown,
especially towards extremities ; femora of hind-pairs have a
deeper hue ; well-defined greenish rings on third pair ; perceptible greenish and brownish annulations on their tibial
and metatarsal joints ; about 16 spines on femora of first pair;
9 or 10 on second ; few spines on thighs of third pair ; spines
on patellary joints ; spine armature of tibiæ and metatarsi
somewhat sparse and irregular ; rather sparsely furnished with
lightish hairs.

Palpi and legs concolorous ; green annulations on digital
joint, which shades off to a brownish colour ; hairs yellowish,
bristle-like spines yellowish, base dark.

Falces light, glossy, yellow-brown ; conical, vertical ; equal
pars digitalis of palpus in length, and femoral joint of a second
leg in stoutness.

Maxillæ brownish-yellow, clouded with olive-brown ; nearly
as broad as long, roundly pointed, moderately inclined over
labium, which is olive-brown, greenish-yellow margins ; rather
wider than long, roundly pointed.

Sternum chocolate-brown, well-defined yellowish lanceolate figure; cordate, moderately wide; eminences opposite coxæ prominent.

Abdomen somewhat depressed above, fore-part broad, rounded, tapers moderately to spinners; ground-colour dull-green, suffused with creamy, greenish-tinted, purple-spotted, lobate flecks, more or less confluent; margins of folium not sharply defined, acute-crenate, tips black, area lightly suffused with brown, except yellow streak on dorsal line, purple spots on flecks darker shade than marginal; a broad, undulating, creamy-coloured band, with dark-(Hooker's)-green margins, connects humeral prominences, curves somewhat round them, terminates with spots of its own colour; appended by a short petiole, springing from its strongly-constricted centre, is a small reniform mark of similar coloration; 4 impressed spots form a trapezoid narrowest in front; on base, which has a deeper-green shade, are two broken wreaths, composed of rather large, pale creamy-green, purple-margined flecks, enclosing a segment of a circle; greenish lateral borders pass into a purple spotted light-brown on ventral surface; shield greenish-brown, margins undulating, creamy-colour, purple spots. *Vulva* yellowish-brown, side-lobes of scapus dark-brown; corpus moderately convex, rapidly compressed into a large, wide, tapering, backward-curved scapus, whose length about equals the greater breadth of corpus vulvæ; whole field displays somewhat narrow transverse wrinkles; free end, or stylus, shows about two wrinkles of normal width, terminates with a conical membranous cap, about twice as broad as long; moderately-deep obtuse notch on upper margin; the broad, tumid margins of scapus show slight longitudinal convolutions.

Mas. — Ceph.-th., long, 3·2; wide, 2·5. Abd., long, 4; wide, 3. Legs, 1, 2, 4, 3 = 14·5, 13·5, 9, 7·3 mm.

Cephalothorax brownish-orange, clouded about caput and sides with lake-brown; hairs yellowish, somewhat sparse; broad-oval; cephalic part flatly convex, eye-eminence moderately prominent; *clypeus* scarcely equals diameter of a fore-central eye; pars thoracica convex, fovea longitudinal, deep; radial and caput striæ fairly defined.

Eyes on dark rings, fore- and hind-row recurved, posterior centrals perceptibly less than an eye's breadth from each other, removed by rather more than that interval from anterior median pair, more than their space from laterals; fore-centrals divided from each other and side-eyes by an interval equal to their diameter and a half; laterals rather more than half size of median eyes, posited obliquely, two-thirds their breadth apart. on low separate tubercular eminences.

Legs, femora of two first pairs, basal third brownish-yellow, fore-end orange-brown, clouded with lake-brown; a broad, broken, brown ring intersects the light and darker shade; patellar joints have the deeper colour of femora; tibiæ yellowish, green tinge, fore-third normal brown; metatarsi yellowish, faint, wide, greenish central and apical annuli; tarsi yellowish, passing into a darker hue; femora of hind-pairs have a deep red-mahogany shade; broken ring on third pair is separated from darker parts by a light streak; patellæ normal hue; tibial joints have two not well-defined reddish-brown annulations on fore-half; annuli on metatarsal joints reddish; legs moderately stout; hairs sparse, yellowish; femora of first and second pairs, 12 or 13 moderately strong black spines beneath, 8 or 9 above; 4 or 5 small spines on patellæ; tibial joints of first, about 13 irregular spines; second pair less; metatarsi about 7; femora of fourth pair, 5 or 6 beneath on fore-end; of third pair, one spine; spine armature of other joints does not differ greatly from first pairs.

Palpi, humeral joint yellowish; increases somewhat in width forwards, scarcely twice length of cubital + radial joints together; cubital joint yellowish, broad-oval, squarely truncated in front, projects two long bristles; radial similar tinge, short, convex on inner side, produced on outer into a long, broad process, acutely truncated on posterior side, few coarse hairs; lamina brownish-yellow, clouded; moderately hairy; placed rather beneath bulb, base developed into a stout conical projection, directed inwards, curving somewhat upwards, in length nearly equals distance from its own base to apex of lamina, more than half as broad as long; base of lamina prolonged on outer side into a large red-brown apophysis, directed backwards, curved forwards, concave in front; most remarkable appendages of genital bulb—suspended near base is a somewhat quadrate, dark membrane, concave in front, inner angle prolonged into a strong tapering apophysis, whose length about equals breadth of membrane at base; next to the latter appendage is a yellowish, oval, vertical organ, convex behind; directly in front, between two dark, backward-curved, stout processes, is a yellowish membrane, convex on anterior face, rapidly constricted into a fine apophysis curving backwards to the above-mentioned vertical organ.

Falces reddish-brown; conical, vertical, as stout as the femur of a fourth leg.

Maxillæ brownish-yellow, base clouded; nearly as broad as long, roundly pointed.

Labium, apex pale, base dark-brown; nearly as long as wide, somewhat pointed, transverse groove forms a collar-like base.

Sternum greenish-yellow, clouded with brown, centre mark; cordate, clathrate.

Abdomen triangular, base rounded, depressed above; integument pinkish creamy-colour, lightly suffused with dull-yellow, numerous small lake spots; creamy-white—border of band dark on posterior side—moderately wide band connects the slightly-developed humeral protuberances; superior surface of base has an olive-green tinge, clouded with black-brown; a broken, recurved wreath of lightish flecks connects apices of transverse band; folium olive-green tinge, extends from band to spinners, serrated, blackish marks on posterior side of serrations; blackish transverse patch below first serration; oblique brownish streaks on lateral margins; lake spots on ventral surface more pronounced, shield somewhat heart-shaped, olive-brown, margin dark.

Taken in the forest near Stratford, *A. T. U.*

Epeira lævigata, sp. nov. Plate XXI., fig. 6.

Femina.—Ceph.-th., long, 3; broad, 2. Abd., long, 5; broad, 4. Legs, 1, 2, 4, 3 = 10, 8·5, 8, 5·8 mm.

Cephalothorax yellowish, green tinge; hairs light, somewhat sparse; pars cephalica moderately convex, roundly truncated, eye-prominence low, lateral index equal to rather more than two-thirds facial; depth of *clypeus* equal to nearly the diameter of a fore-central eye; pars thoracica convex, sides tolerably rounded; fovea suboval, rather deep; normal grooves well defined; profile-contour rises from stalk at an angle of 40 , forward inclination slight.

Eyes on dark rings, hind-centrals lake-coloured; posterior only sensibly recurved, centre eyes largest of eight, rather more distant from each other than they are from fore-centrals, an interval scarcely equalling their diameter; about their space and one-fourth from laterals; anterior row recurved, median eyes dark, rather more than half size of hind-pair, separated by scarcely an eye's breadth and one-half; divided from side-eyes by an interval somewhat shorter than their space; laterals about one-third smaller than anterior centre eyes, posited obliquely, rather more than one-fourth an eye's breadth apart, on very low elevations.

Falces yellowish, suffused with green; conical, vertical, equal in length to digital joint of palpus, and in stoutness to the femur of a second leg.

Maxillæ yellowish, base clouded with olive-brown; dilated forwards, pointed, length somewhat surpasses breadth, inclined towards each other.

Labium greenish-yellow, base dark olive-brown; oval, rather wider than long; more than half length of maxillæ.

Sternum chocolate-brown, median stripe greenish-yellow, bisected by an interrupted yellow streak ; cordate ; eminences in front of coxal joints.

Legs yellowish, green tinge ; faint reddish-brown central and apical annulations on tibiæ + metatarsi ; indication of distal rings on femora ; legs tolerably strong ; hairs yellowish, somewhat sparse ; spines black, tolerably strong and numerous, somewhat irregular.

Palpi greenish-yellow ; hairs light ; black bristle-like spines.

Abdomen ovate, depressedly convex ; broad-conical hump projects over base of cephalothorax : very sparingly clothed with short hairs, few coarse black ; bright pea-green, deepening in tone on lateral margins ; dorsal mark yellowish ; four dark impressed spots form a trapezoid, narrowest in front ; interval between hind-pair and spinners occupied by a wide band of nearly uniform width, somewhat enlarged and rounded forwards ; bordered by a narrow greenish-brown line ; an arcuate line of similar colour limits the first quarter ; median streak lighter shade, bifurcates at cross-line, enclosing a narrow streak to anus. Ventral shield greenish-brown, displays four yellow spots ; elongate. *Corpus vulvæ* greenish-yellow ; represents a large, transversely-wrinkled, subtriangular, curved scapus, rather wider at base than long ; lateral margins not wrinkled, rather broad, beaded, project upwards and outwards ; the dark tapering extremities are sharply bent backwards to base of stylus ; latter organ stout, of uniform breadth, somewhat depressed or grooved above, scarcely one-half length of scapus vulvæ, projects nearly one-third its own length beyond lateral wings of scapus.

Single specimen, forest near Stratford, *A. T. U.*

Sub-Fam. TETRAGNATHINÆ.

Gen. TETRAGNATHA, Latr.

Tetragnatha arborea, sp. nov.   Plate XXI., fig. 9.

Mas.—Ceph.-th., long, 4 ; broad, 3 ; facial index, 1·6. Abd., long, 5 ; broad, 3·4. Legs, 1, 2, 4, 3 = 32, 20, 14·5, 12 mm.

Cephalothorax pale brownish-yellow, red-mahogany colouring on ocular area shades off into a large lanceolate figure, marbled with brown ; base of figure is somewhat sharply and roundly dilated, tapers moderately, compressed apex slopes into thoracic indentation ; lateral margins of caput clouded with reddish-brown ; hairs fine, sparse ; pars cephalica somewhat flatly convex, roundly truncated, lateral index equals space of hind-row of eyes ; pars thoracica convex, prominent above tapering extremity of caput, somewhat depressed behind, sides well rounded : fovea broad-oval, deep ; radial striæ well

defined; profile-contour rises from thoracic junction at an
angle of 25°, slightly curved across cephalic part; *clypeus* perceptibly retreating, height equals two-thirds of a fore-central
eye.

Eyes of fair size, seated on dark-brown oval spots; posterior row sensibly procurved, eyes about equidistant, centre
pair rather more than their radius from each other; anterior
row moderately recurved; median pair removed from one
another by an interval slightly exceeding their radius, and from
side-eyes by scarcely that space, form with hind-pair an oblong
figure rather longer than wide in front; laterals about one-
fourth the smallest of eight, posited obliquely, less than their
radius apart, on low, dark, tubercular eminences.

Legs light-yellow, green tinge, shading off forwards, especially on fore-pairs, to an orange-yellow; coxæ of posterior
pair marked with a large olive-brown spot; three more or less
pronounced red-brown or olive-brown annuli on femora;
patellary joints similar tinge; tibiæ three annulations, central
and apical very broad, less so on hind-pairs; metatarsi of fore-
pairs have the deep colour of annuli; central and terminal
rings on metatarsal joints of hind-pairs; legs moderately
slender; hairs fine, somewhat sparse; spines slender, 15 on
femora of fore-legs, less on other pairs; patellæ project a
strong spine; tibiæ of first, 9; of second, third, and fourth,
about 7; metatarsal joints, about 5 spines.

Palpi light brownish-yellow; slender, length 6mm.; humeral joint rather shorter than cubital + radial together,
perceptibly incrassated; pars cubitalis about one-half length
of penultimate joint, strongly convex above; bristle rather
short; pars radialis tapers moderately to base, projects a short
process above, on inner side; two faint rings; digital joint
scarcely as long as radial, bulbus genitalis complex; lamina
light mahogany-brown; hairs fine, rather sparse; somewhat
oval, obliquely truncated, projects rather beneath bulb; viewed
from outer side, genital bulb displays at base a large, reddish-
brown, linear, membranous process, twice as long as wide,
apex truncated, emarginate, convex and hairy on outer side,
directed forwards; above and attached to former process is a
subreniform projection, apex acutely prolonged; bulb spiral
(plane) yellow-mahogany colour; projecting from its truncated
extremity is a remarkable yellowish, triangular, membranous
process; curving forwards and upwards from beneath latter
organ is a stout lake-black apophysis, which, viewed from front,
shows a rather stout lateral process; perceptible above truncated apex of lamina, seen from inner side, is a low dark ridge,
posterior end produced into an acute process; base of bulb,
above lamina, shows an ovate depression; apex produced into
a free, tapering, forward-directed appendage.

Falces red-mahogany colour; project at an angle of 20°,
somewhat tumid at base in front, fore-third directed upwards
and outwards, about equal in breadth to the femur, and in
length to the tarsus of a leg of first pair: 3 teeth on outer
row.

Maxillæ chocolate-brown, apices light; long, taper somewhat to base, obliquely truncated on inferior side, superior
side rounded, moderately dilated, rather inclined over *labium*,
which is subquadrate, perceptibly curved outwards, less than
one-half length of maxillæ, transverse indentation; colour of
maxillæ.

Sternum yellowish; few black hairs : cordate, eminences
opposite coxæ.

Abdomen oviform, flatly convex above ; hairs pale-yellow,
somewhat sparse; ground-colour pale olive-green, suffused
with moderate-sized, creamy-coloured, lake-spotted flecks ;
integument more or less stained with lake ; folium lyrate, fore-
half broad, margins undulating, dark-brown ; posterior half
tapers to spinners, compressed to about one-half width ; of a
slightly deeper shade and less flecked than lateral margins ;
between and projecting somewhat beyond the four reddish-
lake impressed spots is a hastate figure, apex directed backwards, haft prolonged round base, free from flecks, creamy-
white border ; dorsal field displays 8 dark-brown, transverse
marks, two anterior pairs subtriangular, placed on posterior
side of last pair of impressed dots ; second pair partially
encloses apex of hastate figure ; in centre of third bar—which,
like the fourth and fifth, is drawn out, apices curved, to
margin of folium—is a creamy spot ; two or three posterior
bars black-brown, or reddish, more or less obliterated in
different examples ; lateral margins longitudinally streaked
with black-brown dashes ; ventral shield defined—except at
fore-end—by a broad margin of pale-yellow metallic flecks
showing small lake spots ; centre of shield displays few flecks
and brown stains.

Femina. — Ceph.-th., long, 4 ; wide, 3. Abd., long, 6 ;
wide, 5.   Legs, 1, 2, 4, 3 = 20, 14, 13·2, 8 mm.

Cephalothorax reddish amber-colour, ocular area reddish
dorsal figure mottled with brown at base, breadth of hind-row
of eyes, fore-third incurved, tapers from end of curves, moderately to limit of caput, apex acute, dips into fovea ; displays
two large yellowish spots in line with its greater breadth ;
brownish stains on lateral margins of caput ; hairs light,
sparse ; pars cephalica somewhat aplanate above, roundly
truncated ; lateral index equals breadth of hind-row of eyes,
facial scarcely one-half width of thorax ; *clypeus* moderately
retreating, depth visibly shorter than breadth of a fore-central

eye; pars thoracica broad-oval, prominent at limit of cephalic part; posterior incline depressed; fovea subcircular, deep; radial striæ ill-defined; profile-line rises at an angle of 40°, slopes with a slight curve across caput.

Eyes on dark oval spots, subequal, form an oval figure, anterior row more distinctly curved, centrals posited on slight tubercles, rather less than an eye's breadth from each other and laterals next to them; posterior row divided by nearly equal intervals, exceeding radius of a median eye; four centre eyes form an oblong figure rather longer than wide in front; laterals rather smallest of eight, seated obliquely on low tubercular eminences about their radius from one another.

Legs colour of cephalothorax, faint-green tinge; femora three more or less faint and broken annuli; basal half of patellar joint shaded; tibiæ of first pairs four annulations, hind-pairs three; two on metatarsi; moderately slender; armature sparse, fine outstanding hairs; few bristle-like spines on thighs; single spine near base of metatarsi; two spines beneath tibiæ of fourth; superior tarsal claws—first pair, base straight, evenly and moderately curved forwards, 18 short open teeth; inferior claw sharply bent, free end long, 2 close teeth.

Palpi colour and armature of legs; moderately slender, fully surpass cephalothorax in length; palpal claw moderately curved, free end more than one-half length of claw; 8 rather even comb-teeth.

Falces mahogany-brown; subconical, project at base in front, vertical, inclined outwards; 3 teeth in outer row.

Maxillæ light chocolate-brown, pale apices; about twice as long as broad, taper somewhat to base, inner side obtusely truncated, truncation perceptibly rounded; outer rather dilated.

Labium colour of maxillæ; oblong, nearly twice as wide as long; transversely grooved.

Sternum yellowish; cordate; eminences opposite coxal joints.

Abdomen oviform; ground-colour dark-stone, passing into a dull olive-green, irregularly suffused with lake; flecks numerous, yellow-stone dotted with lake; pattern closely resembles male's; ventral shield lighter shade, flecks golden; centrally marked with brown spots. *Corpus vulvæ* reddish-mahogany; triangular, flatly convex; rather beneath the roundly-truncated apex, which projects slightly over the rima genitalis, is a long narrow-oval orifice.

Male and female examples were rather numerous, more especially in the vicinity of watercourses. Stratford, Taranaki, *A. T. U.*

Tetragnatha multi-punctata, sp. nov.

Femina.—Ceph.-th., long, 2·8; broad, 1·8. Abd., long, 6; broad, 2·3. Legs, 1, 2, 4, 3 = 19·5, 13·5, 10·5, 5·5 mm.

Cephalothorax brownish-yellow, fovea and vein-like lines on caput brown; pars cephalica depressedly convex, roundly truncated, facial index surpasses lateral by one-fourth; depth of *clypeus* scarcely equals diameter of a fore-central eye; pars thoracica oval, depressed; fovea subovate, rather small and shallow; caput and radial striæ moderately defined; profile-line slopes moderately backwards from hind-row of eyes, slightly arched across thorax.

Fore and hind-row of *eyes* somewhat evenly recurved; posterior eyes nearly equidistant; median pair on black oval spots, further from one another than they are from anterior centrals, a space perceptibly exceeding an eye's breadth; fore-centrals posited on dark, lake-tinged rings, fully equal or slightly surpass hind-pair, with whom they form a quadri-lateral figure broader than long, rather closer to each other than they are to side-eyes; laterals seated on strong black tubercles, rather less distant than are the fore- and hind-centre eyes; posterior eye exceeds anterior by about one-third.

Falces light amber-colour, fangs reddish; project at an angle of 45°, divergent; 6 teeth in outer row, 5 close, sixth tooth as far from fifth as it is from apex of falx; inner row, 7 teeth increasing in size and distance from one another.

Maxillæ reddish-brown; sublinear, fore-third divergent, superior angle dilated.

Labium shade darker; rather longer than broad, sides parallel, roundly pointed, transversely grooved.

Sternum yellowish-amber, sides clouded with brown; cordate.

Legs deep straw-colour, green tinge; slender; hairs sparse, fine, outstanding; few long, black, bristle-like spines on all joints except tarsal.

Palpi colour and armature of legs; slender, 3·5mm. long.

Abdomen cylindrical; deep-violet, passing into brown beneath, suffused with silvery lobate flecks; fore-half of dorsal line throws off a series of angular streaks, apices directed forwards, hind-pair straightest. *Corpus vulvæ* yellowish, moderately convex area, viewed from above somewhat campanulate, base contiguous to stigmata; truncated apex shows a long narrow orifice; a wide orange-red process projects from within beyond limit of corpus.

Single example, Taranaki, *A. T. U.*

Tetragnatha flavida, sp. nov.

Mas.—Ceph.-th., long, 3·1; broad, 2. Abd., long, 6; broad, 2. Legs 1, 2, 4, 3 = 20·5, 16, 15·5, 8 mm.

Cephalothorax light brownish-yellow, lateral margins and bifurcating band on caput faintly shaded with brown; posterior end of band and fovea more darkly pencilled; pars cephalica flatly convex, roundly truncated; facial index rather surpasses lateral; depth of *clypeus* equals interval between fore-centre eyes; pars thoracica oval, depressed, fovea deep, oval, longitudinal; normal grooves somewhat slight; profile-contour rather abrupt at posterior incline, somewhat level to limit of caput, occiput perceptibly curved, slope across eye-area.

Eyes on oval dark spots; posterior row moderately recurved, nearly equidistant, shortest interval between centrals, a space equal to nearly twice an eye's breadth; latter pair separated from fore-centrals by a somewhat shorter interval; anterior row more distinctly recurved, median pair rather smaller than hind-pair, and somewhat larger than posterior lateral eyes, divided from each other by nearly their diameter and a half, and from side-eyes by rather less than their space; laterals posited obliquely, interval between them about equal to space dividing centre pairs; fore-eye less than half size of posterior.

Falces brownish-yellow; project at an angle of 45°, moderately convex on inferior and concave on superior side, fore-half somewhat tumid, apex pointed, a strong curved tooth-like process projects above near apex; 8 teeth in outer row, 6 rather close teeth, increasing in length; 2, somewhat larger and more widely separated, on fore-half; inner row, 7 somewhat equidistant teeth.

Maxillæ brownish-yellow, clouded; long, sublinear, pointed, superior angle dilated.

Labium brown, light margins; oval, more than half length of maxillæ.

Sternum dull greenish-yellow, clouded with olive-brown; cordate, eminences opposite coxæ.

Legs deep straw-colour, lightly shaded with olive-brown; slender; hairs fine; few bristle-like spines on all joints except tarsal.

Palpi and legs concolorous; coxal joint as long as radial; pars humeralis enlarged forwards, arcuated, equals radial + digital joints together in length; pars cubitalis campanulate, lobed, about one-half length of penultimate article, projects a strong black bristle at apex; radial joint long, tapers to base, where there is a short blunt process on upper side; breadth at extremity nearly equal to half its own length; bilobate on

12

inner side, upper lobe long, oval ; armed with long black hairs ;
digital joint rather longer than radial ; lamina narrow, sublinear, about twice length of bulb ; hairs rather fine, short ;
first accessory lamina lies on superior surface of bulbus, reaches
the fore-edge, elongate, extremity roundly truncated, tapers to
base ; second oval, terminates at base of bulb ; genital bulb
disciform, brownish-yellow, displays two brown rings ; visible
between accessory laminæ is a reddish, pointed, lip-shaped
process projecting backwards from base of bulbus ; on face of
bulbus genitalis are two somewhat intricate spiral apophyses,
extending to apex of lamina ; outer blackish, narrow, tapering,
convex on inner side ; inner apophysis fulvous, membranous,
semi-transparent, entwined and partially fused with second
half of black apophysis, tapers moderately to apex, which is
deeply cleft ; projects a short process at first spiral curve.

Abdomen cylindrical ; folium yellow stone-colour, suffused
with light-yellow metallic flecks ; dorsal line unflecked, throws
off two pairs of oblique streaks ; side-borders tinged with red-
lake ; lateral margins lightly suffused with brown, yellow
flecks ; ventral field displays three longitudinal bands, spots on
side-bands bright golden-colour, centre band suffused with
brown.

Femina.—Ceph.-th., long, 3·5 ; broad, 2. Abd., long, 7 ;
broad, 2·5. Legs, 1, 2, 4, 3 = 21·2, 14, 13, 6·2 mm.

Cephalothorax light brownish-yellow, two brownish streaks
radiate forwards from limit of caput ; hairs light, sparse ; pars
cephalica depressedly convex, roundly truncated, one-half
breadth of thorax ; depth of *clypeus* equal to space dividing
anterior central eyes from hind-pair ; pars thoracica rather
narrow-oval, depressed above, fovea subcircular, moderately
deep ; striæ somewhat shallow ; caput striæ more deeply
grooved beyond junction ; profile-contour rises abruptly from
thoracic junction, runs in a nearly level line to caput-indenta-
tion, visibly arched forwards, curved across eye-area.

Eyes on black spots, represent two somewhat evenly-recurved rows ; hind-centrals rather larger than fore-centre eyes,
latter pair somewhat exceed posterior laterals ; fore-laterals
less than half size of hind-pair ; centrals of posterior row rather
closer to one another than they are to laterals of same row,
separated from fore-centre eyes by an interval slightly exceeding space between themselves ; anterior median pair about
an eye's breadth and a half apart, scarcely their space from
side-eyes.

Falces light brownish-yellow ; project at an angle of about
45°, somewhat straight on outer side, curved on inner ; outer
row of teeth, 5 ; inner, 7.

Maxillæ light brownish-yellow ; length equal to width of

sternum, sublinear, fore-third bent outwards, inferior angle
rounded.

Labium chocolate-brown, pale margin ; oval, one-third
length of maxillæ.

Sternum greenish-fulvous, passing into brown on margins ;
broad-cordate.

Legs deep straw-colour, fore-half of articles shaded with
reddish-brown ; slender ; hairs fine, very sparse ; few bristle-
like black spines on all joints except tarsal.

Palpi straw-colour ; armature of legs ; slender, length equal
to tibia of a second leg.

Abdomen cylindrical ; folium linear-oval ; ground-colour
olive-stone, suffused with pale-yellow flecks ; dorsal band
unflecked, tolerably wide on fore-half, reduced to a fine streak
on posterior half, enlarged at anus, three strongly-arcuated
lines on central third ; broad band on lateral margins suffused
with confluent silvery flecks, border red-brown, centre streak
paler hue ; ventral region olive-brown, closely spotted with
yellow flecks, except on central band, which is of somewhat
uniform width. *Vulva* yellowish, shaded with stone-brown ;
about as broad as long, convex, rather depressed above, somewhat truncated over the rima genitalis, longitudinal orifice.

A fine male example of this species was contained in *Mr.
T. Kirk's* collection, captured at Belmont by *Miss Kirk*. Male
and female specimens were taken by myself on rushes growing
on the sand-hills near Hawera. Examples were taken in the
forest near Stratford, and a female was captured on a bunch of
rushes above the line of scrub on Mount Egmont. This group
of *Tetragnatha* with cylindriform abdomens as a rule affect
rushes, the position assumed by the spider, and its coloration,
assimilating with its surroundings.

Fam. THOMISIDÆ.

Sub-Fam. PHILODROMINÆ.

Gen. PHILODROMUS.

Philodromus rubro-frontus, sp. nov.

Mas.—Ceph.-th., long, 2·6 ; broad, 2·3. Abd., long, 3·5 :
broad, 2·3. Legs, 1–2, 3–4 = 9·2, 5·5 mm.

Cephalothorax pea-green, facial region crimson-lake ; almost glabrous ; broad-oval ; cephalic part flatly convex,
somewhat squarely truncated ; *clypeus* transversely rugose,
visibly directed forwards, depth slightly exceeds space between fore-centre eyes ; pars thoracica convex, posterior incline indented, normal grooves shallow ; profile-line rises at an
angle of 50°, moderately inclined across occiput, curved over
eye-area.

Eyes small, of nearly equal size. form two somewhat evenly-recurved rows; posterior row nearly equidistant, centre pair slightly the furthest apart; anterior centrals rather closer to one another than they are to posterior median pair, an interval barely surpassing space between latter pair; laterals about one-third larger than centrals, posited on low tubercular eminences, separated by an interval somewhat shorter than that dividing fore- and hind-pairs.

Falces greenish-orange colour, suffused with lake; transversely rugulose; flatly convex, second half linear; slightly inclined forwards; bordered their entire length, outer side, by a dark costa; falx rather more than half as broad at base as long, equal to breadth of anterior row of eyes in length.

Maxillæ spathulate, superior side of second half dilated, directed towards each other.

Labium conoid, two-thirds length of maxillæ; organs yellowish-green, suffused with lake; transversely rugulose.

Sternum yellow-tinted pea-green; few hairs; cordate.

Legs yellowish pea-green, tinged with lake; two first and two hind-pairs of about equal length and strength; hairs, bristles, and spines sparse, latter only on femoral, tibial, and metatarsal joints.

Palpi greenish, suffused with lake; pars humeralis somewhat enlarged forwards, nearly twice length of cubital joint, latter article rather dilated, projects a strong bristle from extremity; radial joint deeply tinged with lake; lower margin developed into a rounded lobe; prolonged forwards on outer side into a black-tipped, pointed, ear-shaped process, nearly one-half length of lamina; digital joint rather longer than two former articles; lamina greenish, suffused with lake; ovate above genital bulb, latter yellowish-brown; subaplanate face displays a red-brown (plane) spiral apophysis, which follows margin of bulb, terminating in centre.

Abdomen oviform, depressedly convex; pea-green, pale flecks, chiefly on margins; four impressed spots form a trapezoid, narrowest in front.

Femina.—Ceph.-th., long, 2·8; wide, 2·8. Abd., long, 5; wide, 4. Legs, 1-2, 3-4 = 8·6, 5·6 mm.

Cephalothorax pea-green, lightly suffused with lake, eye-area lake; sparingly clothed with short whitish hairs, few coarse black on caput; broad-ovate, slightly constricted forwards; cephalic part flatly convex, perceptibly rounded; *clypeus* inclined forwards, height nearly equals interval dividing anterior centre eyes; pars thoracica convex; normal grooves faint; profile-contour rises from stalk somewhat abruptly, moderately inclined forwards across occiput, dips over eye-region.

Eyes on narrow yellow rings, form two evenly-recurved rows; posterior eyes of about equal size, centrals little more distant from each other than they are from side-eyes, form with fore-centre eyes a trapezoid widest behind; anterior median pair slightly exceed hind-centrals in size, rather smaller than fore-laterals, divided by an interval fully equal to that which separates them from hind-pair, scarcely half that space from side-eyes; laterals posited obliquely, hind-eye on a low tubercle, fore-eye slightly elevated, rather further from one another than a fore-eye is from the anterior central.

Falces pea-green; vertical, conical, well-developed costa along superior border, breadth equals width of hind-row of eyes, length equal to radial + digital joints of palpus.

Maxillæ greenish-yellow, lake-brown tinge, somewhat enlarged forwards, pointed, rounded on superior side.

Labium deeper tone; conical, more than one-half length of maxillæ.

Sternum greenish, ovate.

Legs greenish-yellow, lake tinge; first and second, third and fourth, of about equal strength; hairs fine, sparse; 3 or 4 bristle-like spines on femora of first pairs, 1 or 2 on hind-pairs; 1 bristle on patellæ; tibia of fore-leg 2 bristles above, 1, 2, 2, 2 spines; metatarsus, 2, 2, 2, 2 beneath, 1 side-spine; tibia of second leg, 2, 2, 2 on inferior aspect; metatarsus, 2, 1, 2, 2; hind-pairs few bristles above.

Palpi green, lake tinge; armature black bristles, white hairs.

Abdomen pea-green, closely dappled with a paler shade; spinners lake; hairs short, sparse; ovate; 4 impressed spots form a trapezoid narrowest in front; hind-pair deep. *Vulva* represents a not very observable elevation : when pressed below discloses the incision of the rima genitalis.

These specimens were captured in the vicinity of the "Hermitage," Mount Cook, by *Mr. H. Suter.* The species has somewhat close affinities to *P. sphæroides;* the male differing from the latter species in the greater breadth of the pars cephalica, the lateral eyes not being posited on rather prominent cup-shaped tubercles, and in the more ovate form of the abdomen. The absence of eye-prominences, and the more simple form of the vulva, are well-marked differences in the female form. I have given its natural coloration as a uniform pea-green, as being the most probable; but I have recently captured an immature example of *P. anbarus* with an amber-coloured cephalothorax, the normal colour being pea-green.

Fam. LYCOSIDÆ.

Gen. Lycosa, Latr.

Lycosa arenaria, sp. nov.

Femina.—Ceph.-th.. long, 3·4; broad, 2·4. Abd., long, 3·5 ; broad, 2·9.   Legs, 4-1, 2, 3 = 12, 9, 8·5 mm.

Cephalothorax stone-colour, tinged with olive-green, marked with greenish-black speckled transverse stripes, facial region deeper black ; hairs whitish and yellowish, fine, short, adpressed, bristle-like on cephalic part ; length fully equals the patellary + tibial joints of a fourth leg; ovate, slightly compressed forwards ; pars cephalica convex, roundly truncated, lateral index equals two-thirds facial ; height of *clypeus* slightly surpasses space between anterior centre eyes ; pars thoracica convex, border-hem moderately prominent ; indentation longitudinal ; caput and radial striæ fairly well defined ; contour of profile visibly curved across fore-end of caput, moderately inclined to fovea, dips more abruptly to thoracic junction.

First row of *eyes* slightly procurved, much the smallest of eight, centrals about one-third larger than laterals, separated from each other by an interval perceptibly exceeding their breadth, and from laterals by rather more than their radius ; side-eyes rather closer to second row than they are to margin of clypeus ; eyes of second row third larger than posterior pair, rather more distant from them than they are from one another, a space equalling an eye's diameter and a quarter ; posterior eyes one-third further from each other than they are from second line.

Falces fulvous, lightly clouded ; moderately armed with white hairs ; conical, base gibbous in front, divergent, perceptibly inclined forwards, in length scarcely equal to radial + digital joints of palpus, stouter than the femur of first leg ; double row of 3 close teeth. ·

Maxillæ fulvous, somewhat clouded ; base slender, enlarged forwards, rounded slightly, inclined towards *labium*, latter organ olive-brown, margin pale ; nearly as long as broad, rounded, apex truncated, about half length of maxillæ.

Sternum colour of coxæ ; hairs white, sparse ; suboval, broad.

Legs stone-colour, fuscous annulations ; femoral joints, three rings more or less reduced to spots, especially on two first pairs ; patellæ one ring ; tibiæ, basal and central annulations ; metatarsi have three somewhat obliterated rings ; moderately clothed with outstanding white and black hairs ; spines tolerably numerous on all joints except tarsal.

Palpi colour and armature of legs ; pars humeralis fully equals cubital + radial joints together in length ; digital joint one-fourth shorter than two latter articles.

Abdomen oviform ; stone-colour, spotted with black-brown ; thickly clothed with white and yellowish hairs, representing, with dark spots, a well-defined tabby pattern ; ventral region light-brown, hairs white, thick. *Corpus vulvæ*, posterior part fuscous, lighter tone in front ; represents a large transverse oval orifice, dark lateral margins connected on superior side by a fulvous, brown-bordered, membranous costa ; within orifice is a semicircular fulvous lobe, convexity directed towards and connected by a moderately prominent ridge with inferior margin.

Mr. T. Kirk, F.L.S., to whom I am indebted for the specimens, says that these little spiders are hardly visible when at rest on the sand amongst small stones.

Fam. CTENIDÆ.

Gen. CYCLOCTENUS, Koch.

Cycloctenus pulcher, sp. nov.

Femina.—Ceph.-th., long, 4·5 ; broad, 4. Abd., long, 6 ; broad, 4. Legs, of equal length, 15·5 mm.

Cephalothorax brownish - yellow, approximating to drab about margins, ocular area and markings fuscous ; two streaks on either side of cephalic region ; thoracic radii lighter tone, somewhat confluent, limited by the submarginal band ; wedge-shaped contiguous figures round border-hem ; hairs yellowish, short, sparse ; cephalothorax slightly longer than tibial joint ; pars cephalica depressedly convex, sides abrupt, roundly truncated, breadth equal to one-half of thorax ; clypeus scarcely equals space dividing fore-central eyes ; pars thoracica somewhat dome-shaped, rises very perceptibly above plane of occiput ; thoracic groove reddish, longitudinal ; striæ somewhat shallow ; contour of profile rises from the stalk at an angle of 45°, rounded above thorax, slopes slightly across caput to second row of eyes, dips abruptly to margin of clypeus.

Eyes light-brown, except small laterals, which have a pearl-grey lustre ; first row straight, eyes about one-half size of centrals of second line, rather more distant from latter pair than they are from each other, an interval about equal to their diameter ; second row perceptibly procurved ; median pair slightly elevated, an eye's breadth apart, rather less than their radius from laterals, which have a broad - oval form, much the smallest of eight, posited at base of tubercles, little more distant from posterior eyes than they are from centrals of their own row ; hind-pair of eyes do not differ perceptibly in size from median pair of second line, seated obliquely— directed somewhat backwards—on well-developed tubercular eminences, separated by an interval surpassing space between centre pair of next row by one-third.

Legs shade lighter than cephalothorax, fuscous markings; femora approach to drab, spotted, annulations somewhat crenate, interrupted; single ring on patellæ; broken central and distal annuli on tibial + metatarsal joints; hairs somewhat sparse; short spines on superior aspect of femora; 1 spine on patellary joints; tibiæ, 2, 2, 2, 2, 2 beneath, 2 lateral spines; metatarsi, 2, 2, 2, 2 on inferior side; spines on hind pairs nearly as numerous, somewhat irregular.

Palpi deeper tone than legs, annuli well-defined; length, 5·5 mm.; sparingly armed with hairs; long spines on pars digitalis.

Falces rich red-mahogany colour; transversely rugulose; strong fringe along outer margin of groove, otherwise sparingly furnished with hairs; conical, somewhat gibbous at base in front, divergent, directed moderately forwards; stout, length equal to digital joint of palpus; 4 small teeth in superior row, increasing in strength; interspace between fore-tooth and fang rather exceeds length of row; inner side shows 2 somewhat stronger teeth in advance of outer row.

Maxillæ yellowish-brown, clouded; about half as wide at base as long, superior side of second half roundly dilated; inferior margin perceptibly curved, apices emarginate; superior angle displays a dense fringe of hairs; inclined towards *labium*, which has a deeper shade; conical, apex abscinded, concave, three-fourths length of maxillæ.

Sternum deep-fulvous, fuscous clouds round margins; broad-cordate.

Abdomen oviform, margins wrinkled; moderately clothed with short, adpressed, deep-yellowish hairs; brownish-yellow, passing into black-brown on lateral margins and posterior third; fore-part of yellowish region somewhat quadrate, border sinuating, encloses a light blackish-brown oval figure dilated laterally in front; posterior part spotted. Ventral region displays a chestnut-brown lozenge-shaped shield, margins approximating to orange-brown. *Corpus vulvæ* lake-black, passing into a more pronounced lake on margins; large, transverse, broad-oval, moderately convex, superior border shortly prolonged over the rima genitalis, bounded by ridge-shaped costæ terminating abruptly close to the tumid, elliptical, bright-lake-coloured apex.

Wellington, *T. Kirk, F.L.S.*

Fam. ATTIDÆ.

Gen. ATTUS, Walck.

Attus montinus, sp. nov.

Femina.—Ceph.-th., long, 2 ; wide, 1·5. Abd., long, 3 ; wide, 2. Legs, 4, 1, 2, 3. Leg of 4th pair, 4·1 mm.

Cephalothorax reddish-mahogany, passing into dark mahogany-brown on margins; eyes on black spots; hairs yellowish,
short, sparse, chiefly about frontal region, moderate fringe on
clypeus; pars cephalica plano-convex, perceptibly rounded,
sides abrupt, limited by an indentation; height of *clypeus*
equals space dividing centre eyes; thoracic part one-third
longer than cephalic, convex, sides slightly dilated; contour
of profile rises at an angle of 45°, level to posterior eyes,
moderately inclined across ocular area.

Anterior row of *eyes* slightly curved, nearly equidistant,
laterals one-third size of centrals, separated from them by
about their radius; posterior eyes do not differ perceptibly in
size from fore-laterals, posited a little closer to one another,
divided by a somewhat greater interval than that which separates them from lateral border; eyes of second row equidistant; ocular area about one-third broader than long.

Falces red-mahogany colour, transversely rugulose, vertical,
moderately strong, length slightly surpasses the pars digitalis
of palpus.

Maxillæ fulvous, suffused with reddish-brown; nearly
twice as long as wide, second half dilated, rounded.

Labium darker; enlarged at extremity, apex somewhat
rounded, rather less than one-half length of maxillæ.

Sternum chocolate-brown; oval.

Legs, thighs of two first pairs chocolate-brown; patellæ
yellowish-brown; tibiæ chestnut-brown; metatarsi + tarsi
yellow-brown, second half of former articles dark-brown;
posterior pairs fulvous, fuscous basal and distal annulations
on femoral + tibial joints, fore-half of metatarsi similar shade.
First and second legs of nearly equal length, moderately
strong; tibia slightly exceeds patellary joint in length;
metatarsus and tarsus equal; hairs sparse; spine armature
normal.

Palpi bright straw-colour, humeral joint clouded; hairs
whitish, somewhat sparse.

Abdomen oviform, moderately convex above; sparingly
clothed with short whitish and orange-red hairs; ground-
colour drab, perhaps approximating to pale olive-brown;
dorsal band ovoid, earthy-brown, without any determinate
limits, fading away into ground-colour; ill-defined oblique
brown bands on lateral margins; ventral region light-brown.
Corpus vulvæ bright yellow-mahogany colour, two red-brown
wide pointed marks encroach from above across to foveæ;
corpus represents a rather large subcircular, convex, centrally-
depressed area, superior half projects rather outwards over
the rima genitalis, roundly emarginate; within central depression are two circular foveæ, divided by a narrow septum; fus-

cous outer margins of foveæ produced into well-defined triangular processes projecting towards centre of foveæ.
Single specimen, Mount Cook, *H. Suter.*

Attus monticolus, sp. nov.

Femina.—Ceph.-th., long, 2 ; wide, 1·2. Abd., long, 2·5 ;
wide, 1·6. Legs, 4, 1, 2, 3. Leg of 4th pair, 3·8 mm.

Cephalothorax yellowish mahogany-colour, lateral borders
chocolate-brown, deepening in tone round eyes ; dorsum
marked with a brownish, ill-defined, broad hastate figure,
whose apex intersects posterior eyes ; hairs whitish and
orange-red, chiefly on lateral margins of caput, increasing in
length on clypeus ; irides of centre pair of eyes orange-red ;
elevated ; cephalic region plano-convex, widening slightly in
front, sides subvertical ; rather prominent behind posterior
eyes ; limited by a somewhat obovate depression ; height of
clypeus equal to nearly twice diameter of a lateral eye ; sides
of thoracic part very slightly dilated ; profile-contour inclined
forwards, with a perceptible curve, across ocular area, rises
slightly and shortly behind posterior eyes, dips at an angle of
45° to thoracic junction.

Anterior row of *eyes* sensibly recurved, centre pair sub-
touching ; laterals about one-third size of former pair,
separated by an interval equalling one-fourth their breadth ;
posterior pair trifle smaller than fore-laterals, posited rather
closer to each other ; rather less than one-fourth further from
one another than they are from frontal margin ; eyes of second
row somewhat nearer to fore-laterals than they are to dorsal
eyes ; ocular area scarcely one-fourth wider in front than long.

Falces brownish-orange ; somewhat sparse white hairs ;
subconical, vertical, rather slight ; scarcely as long as the
radial + digital joints of palpus.

Maxillæ lighter shade than falces ; arcuated on inferior
side, dilated and rounded at extremity on superior side ;
directed visibly outwards.

Labium light greenish-brown ; conical, somewhat abscinded, less than one-fourth length of maxillæ.

Sternum light-brown ; oval.

Legs brownish-yellow, annulations on femoral, tibial, and
metatarsal joints ; reddish and faint on two first pairs, darker
shade on hind-pairs ; distal rings on metatarsi chocolate-
brown ; first and second pairs do not differ much in length or
strength, moderately strong ; tibiæ cylindrical, slightly surpass genuæ in length ; metatarsal and tarsal joints of about
equal length ; hairs sparse, whitish ; slender spines on femora ;
tibial and metatarsal spines strong.

Palpi straw-colour ; hairs white ; bristles sparse, fine,
black ; palpus rather stout.

Abdomen elongate-oviform; hairs short, very sparse, white and orange-red; deep stone-colour, closely dappled with a paler tone, and few brown dots; dorsum displays on basal fourth a conduplicate mark—apices directed backwards—of a soft black-brown hue; near centre is a wide, procurved, arcuate bar of a deeper shade, intersected by a light-brown sagittate figure, apex directed forwards; posterior fourth shows a dark angular mark with revolute ends; projecting from it towards spinners is a short, somewhat lanceolate figure; inferior half of lateral margins marked with light earthy-brown, close, oblique streaks; ventral region normal colour, border narrow, serrate, interrupted. *Vulva* reddish-brown, approximating to yellow-brown on margins; large, transverse oval eminence, emarginate over the rima genitalis; area occupied by two circular foveæ, intersected by a moderately broad)(-shaped septum; foveæ exhibit within a dark, wide, tapering, revolute membrane.

Captured by *H. Suter*, Mount Cook.

Attus valentulus, sp. nov.

Femina.—Ceph.-th., long, 1·8; wide, 1·8. Abd., long, 3; wide, 2·5. Legs, 4, 3, 1, 2 = 4·1, 4, 3·8, 2·5 mm.

Cephalothorax brown-black; moderately clothed with white and yellow adpressed hairs and erect black; irides orange; elevated; cephalic part aplanate, frontal region slightly rounded, sides abrupt; depth of *clypeus* less than radius of a fore-centre eye; fringe of white hairs; thoracic and cephalic parts of about equal length, sides rounded; profile-line rises somewhat abruptly from thoracic junction, slightly curved to dorsal eyes, moderately inclined forwards across ocular area.

Anterior row of *eyes* sensibly recurved; laterals about one-fourth size of centrals, separated by intervals exceeding radius of a side-eye; dorsal eyes perceptibly larger than laterals, project from moderate prominences, near margin of hind-slope; space between them exceeds interval dividing laterals by nearly one-third; more distant by one-third from each other than they are from frontal margin; eyes of second row posited rather closer to fore-laterals than they are to hind-pair; ocular area one-third broader behind than long.

Falces lake-brown, hairs white; flat, scarcely as long as the pars digitalis of palpus; breadth equals space occupied by fore-centre eyes.

Maxillæ light chocolate-brown, pale apices; enlarged forwards, rounded, inclined over *labium*, latter organ deeper shade than maxillæ, rather more than half their length; oval.

Sternum lake-black; white hairs; small, oval.

Legs brownish-yellow, suffused—except tarsi—with lake-

brown; fore-pair darkest; two hind-pairs of about equal
strength; anterior legs stoutest; thighs compressed; patellæ
+ tibiæ stout, latter articles cylindrical, about one-fifth longer
than patellar joints; metatarsi + tarsi rather slender, latter
articles shortest by one-fourth; white and erect black hairs;
1 short spine on fore-end of femora; tibia of first leg, 2, 2, 2;
metatarsus, 2. 2; second leg, tibia, 2, 2; metatarsus, 2; tibial
joint of third leg, 2 spines at fore-end; metatarsal joints of
hind-pairs, ring of 5 spines; superior tarsal claws—first pair,
basal two-thirds straight, fore-third sharply curved, outer claw
no teeth, inner about 12 or more short close teeth, claw-tuft
strong.

Palpi brown-yellow; sparingly haired; cubital, radial,
and digital joints of about equal length.

Abdomen oviform, moderately convex; brown-black; fairly
clothed with short adpressed yellowish hairs; on basal half
is a bare patch of a lanceolate form, apex directed forwards;
spot of white hairs on lance-head; similar spot on base of
haft, between the latter spot and spinners are three smaller
spots, latter flanked by bare spaces enclosing patches of white
and yellowish hairs; ventral region brownish, hairs moderately
thick. *Vulva* represents two large, yellowish, ovate foveæ,
bordered by brownish beaded costæ, separated by a space
nearly equalling their transverse diameter.

Single example, North Shore, Auckland, *A. T. U.*

Gen. MARPISSA, C. Koch.

Marpissa cineracea, sp. nov.

Femina.—Ceph.-th., long, 2; broad, 1·2. Abd., long, 2·8;
broad, 1·8. Legs, 1, 4, 2, 3. First leg, length 6 mm.

Cephalothorax red-mahogany, passing into olive-brown
about margins, blue-black stains; hairs white and golden,
tolerably thick, adpressed, black erect hairs; irides golden:
clypeus fringe white, long; elongate, cephalic part aplanate,
limited by a transverse indentation; height of *clypeus* equal to
radius of a lateral eye; pars thoracica slightly dilated, rather
more than one-third longer than cephalic part; contour of
profile rises from thoracic junction at an angle of 60°, moderately inclined forwards.

Anterior row of *eyes* visibly recurved, centrals much the
largest, rather nearer to each other than they are to laterals,
a space equal to radius of latter eyes; posterior eyes about
equal to fore-laterals in size, perceptibly closer to them than
they are to each other; eyes of second row posited somewhat
nearer to anterior laterals; ocular area one-third wider than
long.

Falces olive-brown; few white hairs; short, about as long
as radial + digital joints of palpus together, breadth scarcely

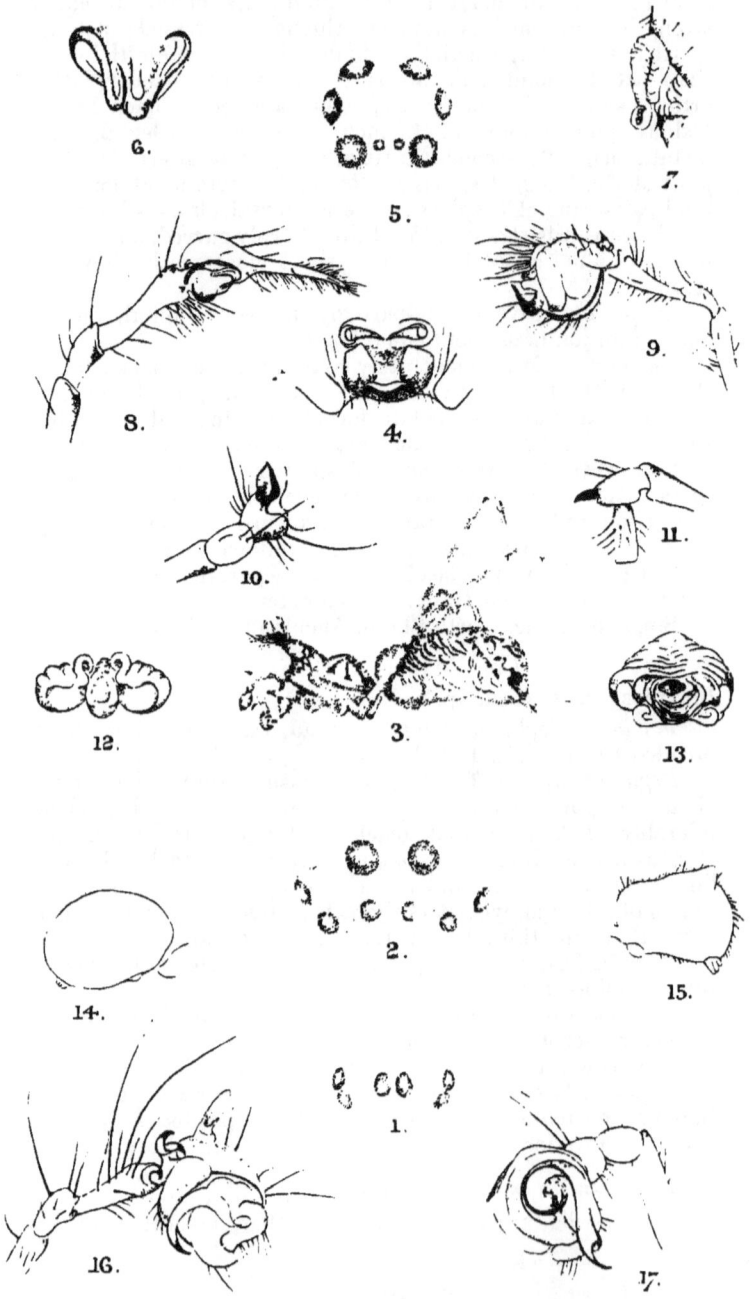

surpasses interval occupied by fore-centre eyes : inwardly inclined.

Maxillæ fulvous ; dilated and roundly truncated at extremity, directed towards each other.

Labium dark-brown, pale tumid margin, broad-oval, less than half length of maxillæ.

Sternum dull olive-brown ; oval.

Legs pale olive-brown, semi-pellucid, faint annulations at articulation of joints ; hairs somewhat sparse, white, and outstanding black ; spines on tibiæ + metatarsi, terminal rings of 5 on metatarsal joints of two hind-pairs ; first and fourth legs of nearly equal length ; femora of first and second pair stout, compressed ; patellæ + tibiæ cylindrical, of nearly equal length, metatarsi + tarsi shorter and more slender than former articles.

Palpi shade paler than legs ; hairs white, few black hairs.

Abdomen elongate-oviform, base truncated, rather depressed above ; moderately clothed with white adpressed hairs ; pale-drab, markings dark-brown ; spotted somewhat in lines ; basal third displays a conduplicate figure, convexity directed forwards, extremities rapidly compressed, revolute ; in centre of dorsal field are two somewhat crescent-shaped figures, inner horns prolonged forwards, nearly confluent ; following the latter is an angular mark, thickening at basal extremities. *Corpus vulvæ* colour of abdomen, foveæ brown ; area occupied by two large, broad-oval, rugose foveæ, divided by a narrow) (-shaped septum.

Stratford, *A. T. U.*

———

EXPLANATION OF PLATE XXI.

Fig. 1. *Oonops septem-cincta*, sp. nov.: Eyes.
Fig. 2. *Habronestes celeripes*, sp. nov.: Eyes.
Fig. 3. *Cornicularia crinifrons*, sp. nov.: Male, eight times natural size.
Fig. 4. *Erycina violacea*, sp. nov.: Maxillæ and lip of female.
Fig. 5. *Habronestes scitula*, sp. nov.: Eyes.
Fig. 6. *Epeira lævigata*, sp. nov.: Vulva.
Fig. 7. *Epeira atri-hastula*, sp. nov.: Vulva.
Fig. 8. *Tegenaria arboricola*, sp. nov.: Palpus.
Fig. 9. *Tetragnatha arborea*, sp. nov.: Palpus.
Fig. 10. *Linyphia pellos*, sp. nov.: Patella and radial joints of palpus.
Fig. 11. *Cornicularia crinifrons*, sp. nov: Patella and radial joints of palpus.
Fig. 12. *Epeira renustula*, sp. nov.: Vulva.
Fig. 13. *Epeira nigro-hastula*, sp. nov.: Vulva.
Fig. 14. *Erycina violacea*, sp. nov.: Profile of female.
Fig. 15. *Linyphia sennio*, sp. nov.: Profile of female.
Fig. 16. *Linyphia sennio*, sp. nov.: Palpus.
Fig. 17. *Erycina violacea*, sp. nov.: Palpus.

www.ingramcontent.com/pod-product-compliance
Lightning Source LLC
Chambersburg PA
CBHW022009190326
41519CB00010B/1453